Improved Indirect Power Control (IDPC) of Wind Energy Conversion Systems (WECS)

Authored by
Fayssal Amrane
LAS Research Laboratory Department of Electrical Engineering, University of Setif-1, Setif, Algeria

&

Azeddine Chaiba
Department of Industrial Engineering, University of Khenchela, Algeria

Improved Indirect Power Control (IDPC) of Wind Energy Conversion Systems (WECS)

Authors: Fayssal Amrane and Azeddine Chaiba

ISBN (Online): 978-981-14-1267-7

ISBN (Print): 978-981-14-1266-0

© 2019, Bentham eBooks imprint.

Published by Bentham Science Publishers Pte. Ltd. Singapore. All Rights Reserved.

First published in 2019.

need for a court order if at any point you breach any terms of this License Agreement. In no event will any delay or failure by Bentham Science Publishers in enforcing your compliance with this License Agreement constitute a waiver of any of its rights.

3. You acknowledge that you have read this License Agreement, and agree to be bound by its terms and conditions. To the extent that any other terms and conditions presented on any website of Bentham Science Publishers conflict with, or are inconsistent with, the terms and conditions set out in this License Agreement, you acknowledge that the terms and conditions set out in this License Agreement shall prevail.

Bentham Science Publishers Pte. Ltd.
80 Robinson Road #02-00
Singapore 068898
Singapore
Email: subscriptions@benthamscience.net

BENTHAM SCIENCE

CONTENTS

FOREWORD i

PREFACE ii
    HOW TO USE THIS BOOK ii
    CONSENT FOR PUBLICATION ii
    CONFLICT OF INTEREST ii
    ACKNOWLEDGEMENTS ii

CHAPTER 1 GENERAL INTRODUCTION 1
    1. INTRODUCTION 1
        2. THE MAIN CONTRIBUTIONS 3
        3. WORK LIMITATIONS 6
    REFERENCES 6

CHAPTER 2 OVERVIEW OF WIND ENERGY CONVERSION SYSTEMS (WECS) 9
    1. WIND POWER DEVELOPMENT 9
    2. WIND TURBINE CONCEPTS 11
        2.1. Fixed Speed Wind Turbines (WT Type A) 11
        2.2. Partial Variable Speed Wind Turbine (VS-WT) Using Variable Rotor Resistance (Type
        B) 12
        2.3. VS-WT Using Partial Scale Power Converter (WT Type C) 12
        A-Advantages of the DFIG [7] 12
        B-Drawbacks of the DFIG [8 - 11] 12
        2.4. VS-WT using Full Scale Power Converter (Type D) 12
    3. CONTROL STRUCTURE OF WTS 13
    4. LITERATURE SURVEY 14
        4.1. Modelling of a WTGS 15
        A- Modelling of DFIG 15
        4.2. Control Strategies for a WT-GS 15
        A-Maximum Power Point Tracking (MPPT) Control 15
        A.1 Intelligent Control 16
        A.2 Other Control Strategies 16
        B-DFIG Control 16
        B.1 Field Oriented Control 16
        B.2 Direct Torque/Power Control (DTC/DPC) 16
        B.3 Adaptive Nonlinear Control (MRAS Observer/MRAC Controller) 17
        B.4 Adaptive Disturbance Rejection Control (ADRC) 17
        B.5 Sliding Mode Control (SMC) 18
        B.6 Backstepping Control (BSC) 18
        B.7 Predictive Direct Power Control (PDPC) and Deadbeat Control 19
        B.8 Input/Output Linearizing and Decoupling Control 19
    NOTES 21
    REFERENCES 21

CHAPTER 3 INDIRECT POWER CONTROL (IDPC) OF DFIG USING CLASSICAL &
ADAPTIVE CONTROLLERS UNDER MPPT STRATEGY 26
    1. INTRODUCTION 26
    2. MATHEMATICAL MODEL OF DFIG 28
    3. CONVENTIONAL INDIRECT POWER CONTROL (IDPC) OF DFIG 30
        3.1. Relationship Between Rotor Voltages and Rotor Currents (Generally Form) 32
        3.2. Relationship Between Stator Power and Rotor Currents 33

3.3. Relationship Between Rotor Voltages and Rotor Currents (Detailed Form) 34
3.4. Synthesis of the Proportional-Integral (PI) Regulator 35
4. WIND TURBINE MATHEMATICAL MODEL 37
4.1. Maximum Power Point Tracking (MPPT) Strategy 39
5. GRID SIDE CONVERTER (GSC) AND DC-LINK VOLTAGE CONTROL [27 - 29] 42
6. ROTOR SIDE CONVERTER (RSC) 44
6.1. Space Vector Modulation (SVM) [31, 32] 46
6.2. LC Filter 52
7. OPERATING PRINCIPLE OF DFIG 56
**8. EXPERIMENTAL RESULTS OF CLASSICAL POWER CONTROL UNDER SUB-
SYNCHRONOUS & SUPER-SYNCHRONOUS OPERATIONS** 59
9. PROPOSED IDPC BASED ON PID CONTROLLERS 59
9.1. Advantages 61
9.2. Drawbacks 61
10. PROPOSED IDPC BASED ON MRAC CONTROLLERS 62
10.1. Definition 62
10.2. Description 62
10.3. Some Mechanisms Causing Variation in Process Dynamics Are 62
10.4. Advantages 62
10.5. Drawbacks 62
11. SIMULATION RESULTS 65
11.1. Mode 1 (Based on PI, PID and MRAC Without MPPT Strategy) 68
11.2. Mode 2 (Based on PI, PID and MRAC with MPPT Strategy- Step Wind Speed) 74
11.3. Mode 3 (Based on PI, PID and MRAC with MPPT Strategy- Random Wind Speed) ... 77
11.4. Robustness Tests12 for Mode 1, Mode 2 & Mode 3 79
CONCLUSION 81
NOTES 82
REFERENCES 82

**CHAPTER 4 A NOVEL IDPC USING SUITABLE CONTROLLERS (ROBUST AND
INTELLIGENT CONTROLLERS)** 86
1. INTRODUCTION 86
2. DRAWBACKS AND PERFORMANCES LIMITATION OF CONVENTIONAL IDPC 88
**3. PROPOSED POWER CONTROL BASED ON TYPE-1 FUZZY LOGIC CONTROL
(T1-FLC)** 89
3.1. Reasons for Choosing Fuzzy Logic 89
3.2. Fuzzy Set Theory and Fuzzy Set Operations 90
3.3. Membership Functions 91
3.4. Mamdani Fuzzy Inference Method 91
A- Fuzzifier 91
B- Knowledge Base 92
C- Inference Engine 92
D- Defuzzifier 92
3.5. MEMBERSHIP FUNCTIONS AND RULE BASE 93
**4. PROPOSED POWER CONTROL BASED ON TYPE-2 FUZZY LOGIC CONTROL
(T2-FLC)** 96
4.1. Overview of Type-2 Fuzzy Logic Controller Toolbox 97
4.2. Design of Type-2 Fuzzy Logic Controller 98
5. PROPOSED POWER CONTROL BASED ON NEURO-FUZZY CONTROL (NFC) 101
5.1. Layer I: Input layer 102
5.2. Layer II: membership layer 103

      5.3. Layer III: rule layer 103
      5.4. Layer IV: output layer 103
  6. SIMULATION RESULTS 105
      6.1. Mode 1 (Based on T1-FLC, T2-FLC 3x0026; NFC, Without MPPT Strategy) 108
          A-Novel IDPC based on T1-FLC: (Fig. 4.16 to the left side): 108
          B-Novel IDPC based on T2-FLC: (Fig. 4.16 to the middle side): 108
          C-Novel IDPC based on NFC: (Fig. 4.16 to the right side): 108
      6.2. Mode 2 (Based on T1-FLC, T2-FLC NFC, with MPPT Strategy- Step Wind Speed) 109
          A- Novel IDPC based on T1-FLC: (Fig. 4.17 to the left side): 109
          B- Novel IDPC based on T2-FLC: (Fig. 4.17 to the middle side): 109
          C- Novel IDPC based on NFC: (Fig. 4.17 to the right side): 109
          B-Novel IDPC based on T2-FLC: (Fig. 4.18 to the middle side): 112
          C-Novel IDPC based on NFC: (Fig. 4.18 to the right side): 112
      6.3. Mode 3 (Based on T1-FLC, T2-FLC NFC, with MPPT Strategy- Random wind Speed) 112
          A-Novel IDPC based on T1-FLC: (Fig. 4.19 to the Left Side): 112
          B-Novel IDPC based on T2-FLC: (Fig. 4.19 to the Middle Side): 112
          C-Novel IDPC based on NFC: (Fig. 4.19 to the Right Side): 114
      6.4. Robustness Tests7 for Mode 1, Mode 2 Mode 3 114
          A-Mode 1 (Novel IDPC based on T1-FLC, T2-FLC NFC): 115
          B-Mode 2 (Novel IDPC based on T1-FLC, T2-FLC NFC): 115
          C-Mode 3 (Novel IDPC based on T1-FLC, T2-FLC NFC): 115
  **7. WIND-SYSTEM PERFORMANCES RECAPITULATION UNDER SIX (06)
PROPOSED IDPC ALGORITHMS** 117
  CONCLUSION 118
  NOTES 118
  REFERENCES 119

CHAPTER 5 GENERAL CONCLUSION 121
      5.1. FUTURE WORKS 122

APPENDIX A: WECS PARAMETERS 123

LIST OF ABBREVIATIONS 126

LIST OF ACRONYMS 128

SUBJECT INDEX 131

FOREWORD

During the past decade, the installed wind power capacity in the world has been increasing more than 30%. Wind energy conversion system (WECSs) based on the doubly-fed induction generator (DFIG) dominated the wind power generations due to the outstanding advantages, including small converters rating around 30% of the generator rating, lower converter cost.

Due to the non-linearity of wind system, the DFIG power control presents a big challenge especially under wind-speed variation and parameter's sensibility.

To overcome these major problems; an improved IDPC (Indirect Power Control); based on PID "Proportional-Integral-Derivative" controller, was proposed instead the conventional one (based on PI), in order to enhance the wind-system performances in terms; power error, tracking power and overshoot. Unfortunately using robustness tests (based on severe DFIG's parameters changement); the wind-system offers non-satisfactory simulation results which were illustrated by the very bad power tracking and very big overshoot (> 50%).

In this context; adaptive, robust & intelligent controllers were proposed to control direct & quadrature currents (Ird & Irq) under MPPT (Maximum Power Point Tracking) strategy to main the unity power factor (PF≈1) by keeping the reactive power at zero level. In this case, the new IDPC based on intelligent controllers offered an excellent wind-system performance especially using robustness tests, which offered a big improvement especially using Type-1 Fuzzy Logic Controller (T1-FLC), Type-2 Fuzzy Logic Control (T2-FLC; is the New class of fuzzy logic) & Neuro-Fuzzy Logic (NFC).

In this sense, I think that this edited book is an important contribution to help students already in mastery of the basis of power electronic circuits and control systems theory to achieve these pedagogical goals. The proposed book describes with easy manner the modeling & control of Wind-turbine DFIG in order to control the stator powers using different topologies of robust, adaptive and intelligent controllers.

The book present numerous intelligent control techniques that help in the control design of the DFIG wind-system (WT).

The textbook *"Improved Indirect Power Control (IDPC) of Wind Energy Conversion Systems (WECS)"* proposes a collection of concepts, organized in a synergic manner such that to ease comprehension of the WT control design.

The book's contribution goes towards completing the already existing literature by offering a useful integration of control techniques, worthy to be read, understood and employed in the various WT applications.

Please enjoy reading this book.

Dr. Ali CHEBABHI
ICEPS (Intelligent Control & Electrical Power Systems),
Sidi-Bel-Abbes Mohamed El Bachir El Ibrahimi University, BBA,
Algeria

PREFACE

HOW TO USE THIS BOOK

This book offers advanced Power Control such as: Indirect Power Control (IDPC) to overcome wind-system DFIG limitation performances under different wind speed and parameters changement conditions.

This book is addressed to students of: License, Master degrees and also for Post-graduation (PhD students) in order to understand the wind-system basics especially: Power electronics control (*in this proposed Book we used SVM in order to fix the switching frequency*), Power-flow DFIM diagram & Maximum power point tracking strategy.

CONSENT FOR PUBLICATION

Not applicable.

CONFLICT OF INTEREST

The authors declare no conflict of interest, financial or otherwise.

ACKNOWLEDGEMENTS

Firstly, I would like to thank Allah, for His mercy on me during all my life, and praise Prophet MOHAMMAD (Peace be upon him!). I would like to express my appreciation to all those who gave me the possibility to complete this book. I wish to express my best gratitude and thanks to my Co-Editor, Pr Azeddine CHAIBA, for his technical guidance, his intellectual support and encouragement of my research work.

Fayssal Amrane
LAS Research Laboratory Department of Electrical Engineering
University of Setif-1, Setif
Algeria

Azeddine Chaiba
Department of Industrial Engineering
University of Khenchela
Algeria

CHAPTER 1

General Introduction

Abstract: In this chapter, a brief general introduction focuses on the well-known topologies of wind energy conversion systems (WECS), on proposed controls and generators by the scientific researchers. One part will be devoted to the latest research that has addressed the performance problems of wind systems and their results (in simulation). There will be also some arguments that reflect the main proposed ideas in this eBook, the proposed selections and their applications in simulation. We present the selecting criteria in particular the type of: generator, controls and theirs application in simulation studies. Also, we discuss in a detailed section on the different contributions of eBook that define the improvement of the proposed algorithms in each chapter. Furthermore, the organization and structure of eBook will be as follow; chapter one is devoted on the state of the art of wind systems and their controls, in particular using the doubly fed induction machine (DFIM). The simulation part is provided in two chapters (3 and 4). The limitations and problems encountered during the realization of this eBook are well described in the following section. After solving problems, very satisfactory simulation results have been found which reflect the quality of the scientific contribution including more papers of conferences; Journal papers were published during this eBook.

Keywords: Doubly Fed Induction Generator (DFIG), Power Electronics (PEs), Wind Energy Conversion Systems (WECS's), Wind Turbines (WTs).

1. INTRODUCTION

The growing connection of wind turbines has augmented at a quick pace over the last years. Installed wind power production, which is presently higher than 440 GW, is predictable to surpass 760 GW by 2020, creation this form of renewable energy an important element of the current and future energy supply systems [1 - 3]. The wind energy raises more important than any other renewable energy sources and is becoming really a significant factor in the recent energy supply system [4].

In the 1980, the PEs (Power Electronics) WTs (wind turbines) was a soft starter used to primarily interconnect the induction generator with the electrical grid, only thysistors were used and they did not require to carry the power continuously [5].

Fayssal Amrane & Azeddine Chaiba

In the 1990s the PE technology was essentially used for the rotor resistance control of wound-rotor induction generator (WRIG), where further advanced diode bridges with a chopper were used to control the rotor resistance for generator [6], particularly at rated power process to reduce loading and mechanical stress. Since 2000, the bidirectional power flow have introduced with more progressive voltage source converters; the PEs started to handle the produced power from the WTs, first, by partial scale of power capacity for doubly fed induction generators (DFIGs), and then by the full scale of power capacity for asynchronous or synchronous generators (A/SGs) [5, 6].

Although the WTs can be considered into various structures in terms of the generator type, with/without the gearbox, or the rating of the power electronic converter, it is common to divide the WTs system into a partial-scale power converter equipped with a DFIG and a full-scale power converter together with either a synchronous generator (SG) or an induction generator (IG) [7, 8]. Presently, the DFIG system configuration the occupies close to 50% of the wind energy market, due to its small size, light weight, and cost-effectiveness of the generator, as well as the relatively small and economic power converter [9, 10].

The variable-speed WECSs can be worked in the maximum power point tracking (MPPT) mode to extract the maximum energy from wind. For this raison, good-calibrated mechanical sensors, such as encoders and anemometers/ resolvers, are essential in order to obtain the information of wind speed and generator rotor speed/position. But, the usage of mechanical sensors raises the cost, hardware difficulty of WECSs [11, 12]. These difficulties can be resolved by adopting position/speed sensorless control schemes [13].

The DFIG' conventional control approaches are generally based on Field oriented control (FOC) algorithms [14, 15]. In the past few years it suffers from the handicap of the generator parameters changement, which comes to compromise the robustness of the control device. Hence, the regulator should accommodate the effects of uncertainties and maintain the system steady against a big variation of system parameters. The traditional PI-based controllers cannot totally fulfill stability and performance necessities [15]. Their optimal PI's parameters can be defined by other approaches such as genetic algorithm (GA) or particle swarm optimization (PSO) [16 - 18]. Power converter and drive system have inherent features, such as non-linearies, inaccessibility of an accurate model or excessive complexity, that call for intelligent control approaches such as neural networks (NN), fuzzy logic (FL) [19, 20]. The dynamic performance of a WTS can be substantially enhanced by the application of smart methods for the PES control that are used in WPG systems. Hence, the aims of efficient wind power integration in the power system can be successfully accomplied.

Fuzzy logic (FL) has been applied for WPG control [21, 22]. The FL based controller is able to be implant, in the control strategy, the qualitative knowledge of an operator or field engineer about the system, but has been assessed for its limits, such as the lack of a formal design methodology, the difficulty in predicting stability and robustness of FL controlled systems [23]. The artificial neural networks (ANNs) based controllers have been used as these controllers can be formed straight by using the input-output information of the indefinite system, without requirement any previous model structure. However, to choice an ideal structure, parameter values and the number of training sets are still crucial concerns. To take benefit of their strengths and to mitigate their disadvantages, numerous hybrid methods have been planned [24]. A hybrid system can be achieved by, for example, combining a fuzzy inference system and adaptive neural networks (*i.e.*, the adaptive neuro-fuzzy inference system (ANFIS)) [25]. ANFIS based controllers have been successfully implemented for numerous power systems and PE applications [26, 27].

On the other side, the system is greatly nonlinear. Thus, linearization operating point cannot be applied to design the controller. Nonlinear control methods can be used to efficiently solve this problem [28, 29]. In attempt to reach high performances in the steady and stransient states, a diverse nonlinear control configuration must be applied. In the recent years, several modified nonlinear state feedback such as Input-output feedback linearization control (I/OLC), Sliding Mode Control (SMC); Backstepping Mode Control (BMC) and Model Predictive control (MPC) have been applied to more develop the control performances [30].

2. THE MAIN CONTRIBUTIONS

In the review of the DFIG-based wind system in last decade, it can be seen that the majority relies on the regulation of: speed, flux, torque, current and powers. More than 75% of the published articles (mainly based on *"IEEE and Science Direct"* databases between 2005 and 2017) concerning the study and development of the DFIG-based wind system are basically focused on three (03) main controls: vector control (rotor flux and torque), predictive control and direct power control (stator active and reactive power). In this eBook, we are interested in power control (in terms of modeling) whose main objective is to improve the quality of energy transmitted into the network by integrating and developing new algorithms in order to overcome or mitigate drawbacks of conventional controls in transient and steady states during the wind speed variation and under robustness tests.

A detailed simulation study in power control using PI (*Proportional-Integral*) regulators (in order to control the stator powers "P_s and Q_s" and the rotor currents

"I_{rd} and I_{rq}" according to 04 loops respectively) is developed according to three modes, as follows:

- Mode 1: Without the MPPT strategy (imposed power profiles).
- Mode 2: With the MPPT strategy (wind in step form).
- Mode 3: With the MPPT strategy (wind in random form).

(*Knowing that all control algorithms in this eBook are developed using these three modes*).

The MPPT (*maximum power point tracking*) is used in order to extract the maximum power despite the wind speed variation (step or random wind forms) by maintaining the reactive power at zero level means power factor near to the unity.

Some drawbacks appears in simulation studies especially in with/without robustness tests (Knowing that we used the same robustness tests in chapters: 3 and 4) such as:

- An important overshoot is noted (more than + 50%).
- The coupling terms between the parameters of the both axes (d and q) has negative influence on the wind-system performances, especially in high wind-power generation (HWPG).
- The long response time (a visible delay of the measured value relative to that of the reference) order of 10e-2 (sec).
- A bad power tracking of the measured value relative to that of the reference especially if the profile is in the step form.
- Poor power/voltage quality which will be transmitted to the grid; a bad THD that exceeds IEEE standards ($>> + 5\%$).
- A remarkable power error for conventional power control sometimes exceeding 25% of the rated power ($\pm$ 1000 (W) for a rated power of 4 kW).
- The conventional regulators (PI regulators) depend on the DFIG's parameters.

In this context, several approaches have been proposed in order to overcome or minimize these drawbacks (already mentioned above). As a first step, a conventional regulator called PID (Proportional-Integral-Derivate) more developed in term of minimization error and overshoot is proposed instead the PI controllers to control P_s, Q_s, I_{rq} and I_{rd} respectively. Remarkable improved performances are noted for the three modes -Mode: 1, Mode: 2 and Mode: 3; already mentioned above- especially without robustness tests. After applying the 3rd test (green color of the curves/robustness tests section in each chapter) a bad tracking of the active power is particularly apparent when the wind speed varies severely, which means that the PID regulator -for rotor current control: Ird and

Irq- is unable to track the power reference during the sudden wind speed variation and DFIG's parameters variation.

For this reason, intelligent controllers are used to correct the failure of conventional controllers in transient and steady states, such as: "Type-1 Fuzzy Logic Control (T1-FLC)", "T2-FLC (Type-2 fuzzy logic control)" and "NFC (Neuro-fuzzy control)"; knowing that all these proposed controllers (already mentioned above) are used in order to control the d-q axes rotor currents components (I_{rd} and I_{rq}) by keeping PID controllers for stator active and reactive powers tuning (P_s and Q_s). Intelligent regulators have been proposed for the control of "I_{rq} and I_{rd}", in same time; there will be more improved results than used only for "P_s, Q_s, I_{rq} and I_{rd}", the aim of this select is minimizing time computing and in same time maintaining a good performance, then it is to look for a robust controllers which does not depend on DFIG's mathematical model.

In this context, the high-performance regulators known by the name of "intelligent regulators"; are set up to remedy these problems -performances limitation-, three (03) intelligent regulators: T1-FLC, T2-FLC and NFC are used to correct power error especially under robustness tests. T2-FLC and T1-FLC are fuzzy controllers based on the inferences (inputs and outputs in triangular or trapezoid forms) and linguistic rules -depends on the inferences number choosing for studied system to the power of number of inputs, exp: 7 inferences in triangular form "for inputs and output respectively" and 2 inputs; means: $7^2 = 49$ rules - to initiate the optimal calculation of the desired value, noting that the computational algorithm interface is integrated in *MATLAB®/Simulink* software. Knowing that; T2-FLC controller based on three (03) dimensions more than T1-FLC (only two dimensions), this difference in dimensional form generates a complexity of mathematical model of the controller itself and aims to minimize error of the desired value known by optimal value despite the parameters variation of the wind-system.

T2-FLC represents the most developed fuzzy family generation in terms of precision and robustness. NFC regulator is a combination between the fuzzy logic strategy and the artificial neural network (ANN) to have theirs qualities at the same time: to remedy the dependence problem of the wind-system mathematical model and to minimize the calculation of the optimal value while maintaining the robustness despite the parametric variation. Excellent results have been found compared to those found for the last proposed controllers which reflect the robustness of the proposed controller.

3. WORK LIMITATIONS

We found several problems and the majority of them were reminted *i.e.*; the Simpower Systems model in Matlab/Simulink takes a long time in simulation; sometimes for several minutes especially if the sampling time is between $1e^{-6}$ (sec) and $1e^{-5}$ (sec), and this can cause problems in the computer (the PC). If the studied system includes a simple algorithm with a few Simulink-blocks (maximum 2 control loops) this does not pose a problem in general, and if the studied system is complicated with several loops -as in the majority of the algorithms of this eBook- the solution is translated in this case by the realization of the blocks based on the mathematical model of the studied system because the Simpower Systems library contains dozens of algorithms in the same block means that; the studied system is near to the real one; exp: DTC control. By using these simulation blocks, the simulation time is minimized to just a few seconds. It is necessary to note that the calculation time of the proposed power algorithm using T2-FLC will be took more time (nearly twice) than T1-FLC; the reason was the complexity of T2-FLC structure (using three (03) dimensions) compared to T1-FLC (based only on two (02) dimensions).

REFERENCES

[1]   F. Blaabjerg, and K. Ma, "Wind Energy Systems", *Proc. IEEE,* vol. 105, no. 11, pp. 2116-2131, 2017.
[http://dx.doi.org/10.1109/JPROC.2017.2695485]

[2]   Ren21, "Renewables 2016: Global Status Report (GSR)", http://www.ren21.net

[3]   Gwec, "Global Wind Statistics 2016", www.gwec.net

[4]   F. Blaabjerg, and K. Ma, "Future on Power Electronics for Wind Turbine Systems", *IEEE J. Emer. Selec. Topcs. Power. Electron.,* vol. 1, no. 3, 2013.
[http://dx.doi.org/10.1109/JESTPE.2013.2275978]

[5]   Z. Chen, J.M. Guerrero, and F. Blaabjerg, "A review of the state of the art of power electronics for wind turbines", *IEEE Trans. Power Electron.,* vol. 24, no. 8, pp. 1859-1875, 2009.
[http://dx.doi.org/10.1109/TPEL.2009.2017082]

[6]   A.D. Hansen, F. Iov, F. Blaabjerg, and L.H. Hansen, "Review of contemporary wind turbine concepts and their market penetration", *J. Wind Eng.,* vol. 28, no. 3, pp. 247-263, 2004.
[http://dx.doi.org/10.1260/0309524041590099]

[7]   V. Yaramasu, B. Wu, P.C. Sen, S. Kouro, and M. Narimani, "High-power wind energy conversion systems: State-of-the-art and emerging technologies", *Proc. IEEE,* vol. 103, no. 5, pp. 740-788, 2015.
[http://dx.doi.org/10.1109/JPROC.2014.2378692]

[8]   D. Zhou, and F. Blaabjerg, "Bandwidth oriented proportional-integral controller design for back-to-back power converters in DFIG wind turbine system", *IET Renew. Power Gener.,* vol. 11, no. 7, pp. 941-951, 2017.
[http://dx.doi.org/10.1049/iet-rpg.2016.0760]

[9]   M. Liserre, R. Cardenas, M. Molinas, and J. Rodriguez, "Overview of multi-MW wind turbines and wind parks", *IEEE Trans. Ind. Electron.,* vol. 58, no. , 4, pp. 1081-1095, 2011.
[http://dx.doi.org/10.1109/TIE.2010.2103910]

[10]   R. Cardenas, R. Pena, S. Alepuz, and G. Asher, "Overview of control systems for the operation of

DFIGs in wind energy applications", *IEEE Trans. Ind. Electron.*, vol. 60, no. 7, pp. 2776-2798, 2013.
[http://dx.doi.org/10.1109/TIE.2013.2243372]

[11]   Y. Zhao, C. Wei, Z. Zhang, and W. Qiao, "A Review on Position/Speed Sensorless Control for Permanent-Magnet Synchronous Machine-Based Wind Energy Conversion Systems", *IEEE J. Emerg. Sel. Top. Power Electron.*, vol. 1, no. 4, pp. 203-216, 2013.
[http://dx.doi.org/10.1109/JESTPE.2013.2280572]

[12]   W. Qiao, X. Yang, and X. Gong, "Wind speed and rotor position sensorless control for direct-drive PMG wind turbines", *IEEE Trans. Ind. Appl.*, vol. 48, no. 1, pp. 3-11, 2012.
[http://dx.doi.org/10.1109/TIA.2011.2175877]

[13]   W. Qiao, "Intelligent mechanical sensorless MPPT control for wind energy systems", *Proc. IEEE Power and Energy Society General Meeting,* 2012 pp. 1-8

[14]   N. Khemiri, A. Khedher, and M.F. Mimoun, "Wind Energy Conversion System using DFIG Controlled by Backstepping and Sliding Mode Strategies", *Int. J. Renew. Energy Res.*, vol. 2, no. 3, pp. 3-11, 2012.

[15]   Á. Luna, K. Lima, P. Rodríguez, E.H. Watanabe, and R. Teodorescu, "Comparison of Power Control Strategies for DFIG Wind Turbines", *Industrial Electronics, IECON 2008. 34th Annual Conference of IEEE,* 2008
[http://dx.doi.org/10.1109/IECON.2008.4758286]

[16]   B. Yang, L. Jiang, L. Wang, W. Yao, and Q.H. Wu, "Nonlinear maximum power point tracking control and modal analysis of DFIG based wind turbine", *Industrial Electronics, Electrical Power and Energy Systems.*, vol. 74, no. xx, pp. 429-436, 2016.
[http://dx.doi.org/10.1016/j.ijepes.2015.07.036]

[17]   M. Rahimi, and M. Parniani, "Dynamic behavior analysis of doubly-fed induction generator wind turbines-the influence of rotor and speed controller parameters", *Int. J. Electr. Power Energy Syst.*, vol. 32, pp. 464-477, 2010.
[http://dx.doi.org/10.1016/j.ijepes.2009.09.017]

[18]   F. Wu, X-p. Zhang, K. Godfrey, and P. Ju, "Small signal stability analysis and optimal control of a wind turbine with doubly fed induction generator", *IET Gener. Transm. Distrib.*, vol. 1, no. 5, pp. 751-760, 2007.
[http://dx.doi.org/10.1049/iet-gtd:20060395]

[19]   G.C.D. Sousa, and B.K. Bose, "Fuzzy logic applications to power electronics and drives – an overview", *Proc. IECON 1995,* 1995pp. 57-62
[http://dx.doi.org/10.1109/IECON.1995.483333]

[20]   Y. Dote, and R.G. Hoft, *Intelligent Control: Power Electronic Systems* Oxford University Press: New-York, 1998.

[21]   M.G. Simoes, B.K. Bose, and R.J. Spiegel, "Design and performance evaluation of a fuzzy-logi--based variable-speed wind generation system", *IEEE Trans. Ind. Appl.*, vol. 33, no. 4, pp. 956-965, 1997.
[http://dx.doi.org/10.1109/28.605737]

[22]   H.M. Soloumah, and N.C. Kar, "Fuzzy logic based vector control of a doubly-fed induction generator for wind power application", *Wind Eng.*, vol. 30, no. 3, pp. 201-224, 2006.
[http://dx.doi.org/10.1260/030952406778606232]

[23]   B. Singh, E. Kyriakides, and S.N. Singh, "Intelligent Control of Grid Connected Unified Doubly-Fed Induction Generator", *IEEE Conference,* 2010
[http://dx.doi.org/10.1109/PES.2010.5589896]

[24]   P. vas, *Artificial-intelligence-based electrical machines and drives* Oxford University Press: New-York, 1999.

[25]   J.h. Roger Jang, "ANFIS: Adaptive-network based fuzzy inference system", *IEEE Trans. Syst. Man*

Cybern., vol. 23, no. 2, pp. 665-685, 1993.
[http://dx.doi.org/10.1109/21.256541]

[26]   P. Tripathy, *"Development of Adaptive Supplementary Feedback Controller for GUPFC"*, Master Thesis (in English Language), IIT Kanpur, India, 2006.

[27]   C. Potter, and M. Negnevitsky, "Very short-term wind forecasting for Tasmanian power generation", *IEEE Trans. Power Syst.,* vol. 21, no. 2, pp. 965-972, 2006.
[http://dx.doi.org/10.1109/TPWRS.2006.873421]

[28]   J. Karthikeyan, S.K. Kummara, C. Nagamani, and G.S. Ilango, "Power control of grid connected Doubly Fed Induction Generator using Adaptive Back Stepping approach", in Proc", *10ᵗʰ IEEE International Conference on Environment and Electrical Engineering EEEIC'2011,* 2011 Rome
[http://dx.doi.org/10.1109/EEEIC.2011.5874758]

[29]   K. Abbaszadeh, and S. Roozbehani, "A New Approach for Maximum Power Extraction from Wind Turbine Driven by Doubly Fed Induction Generator Based on Sliding Mode Control", *Energ. Manag.,* vol. 1, no. 2, 2012.

[30]   A.L. Nemmour, F. Mehazzem, A. Khezzar, and R. Abdessemed, "Advanced Backstepping controller for induction generator using multi-scalar machine model for wind power purposes", *J. Renew. Energy,* vol. 35, no. 10, pp. 2375-2380, 2010.
[http://dx.doi.org/10.1016/j.renene.2010.02.016]

CHAPTER 2

Overview of Wind Energy Conversion Systems (WECS)

Abstract: The aim of this chapter is to present an overview of the state of technology and discuss some technology tendency in the Power Electronics (PE) used for Wind Power Applications (WPA). Firstly, technological and commercial developments in wind power generation are generally discussed. Next, the wind turbine concept is illustrated and explained using different types of generator. The control structure of wind-turbines (WTs) is explained using DFIG, Asynchronous and Synchronous Generator (ASG and SG). Finally, the last section focuses on a detailed literature review describing DFIG based wind turbine-generator systems in terms of modeling and control strategies.

Keywords: A wound rotor induction generator (WRIG), Asynchronous and Synchronous Generator (ASG and SG), Doubly Fed Induction Generator (DFIG), Wind Turbines (WTs), Wind Power Applications (WPAs).

1. WIND POWER DEVELOPMENT

The increasing wind power capacity between 1999 and 2020 is illustrated in Fig. (**2.1**), and it can be shown that the wind power (WP) has developed fast to an ability of 283 GW with nearly 45 GW installed only in 2012, and this number is probable to reach 760 GW in 2020 on reasonable scenario [1 - 3]. WP raises more important than any other renewable energy and is becoming certainly a significant player in the recent energy supply system. For example, Denmark has a high diffusion by WP and today more than 30% of the electric power consumption is enclosed by wind. This state has even the desire to attain 100% non-fossil based power generation system by 2050 [4]. With regard to markets and constructors, the United States has become the largest market with an installed capacity of more than 13.1 GW in 2012, with China (13 GW) and the Europe Union (11.9 GW) sharing about 87% of the world market [4].

Giga Watts

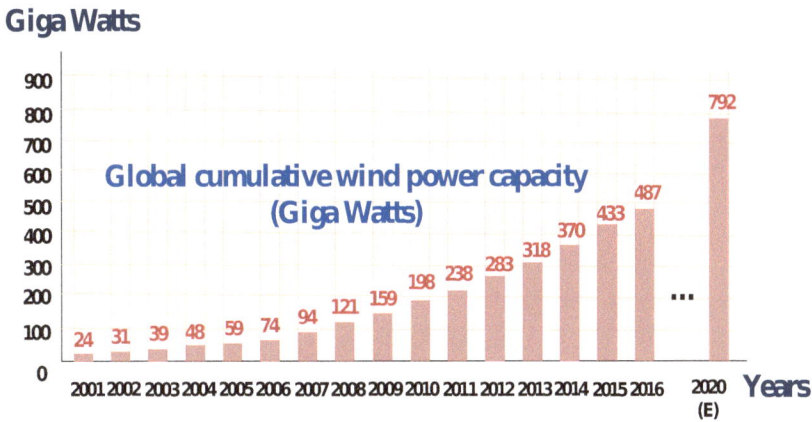

Fig. (2.1). Global increasing wind power capacity from 2001 to 2020 [1].

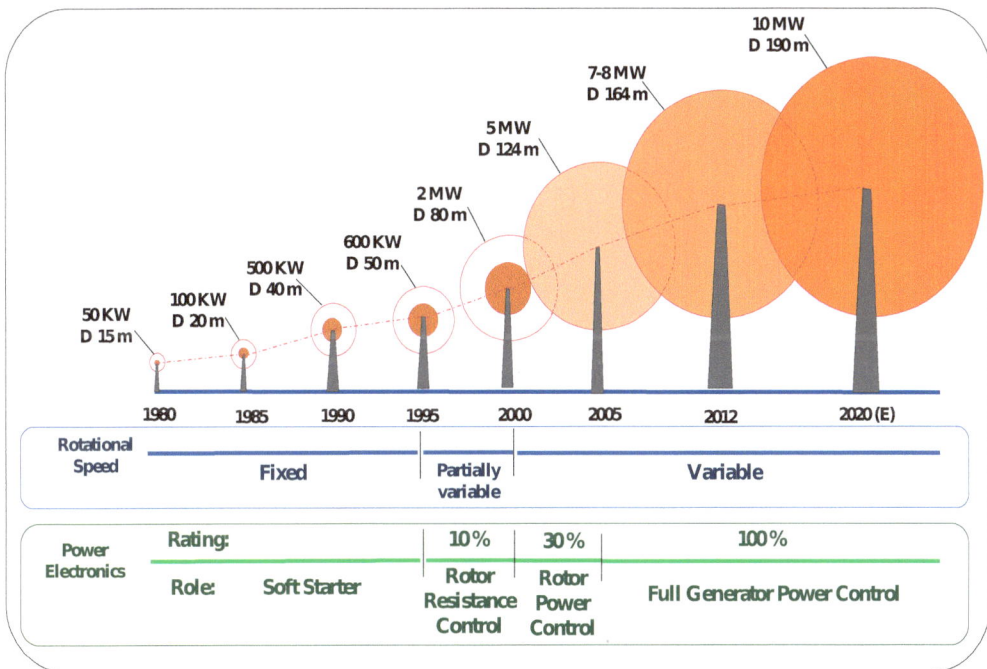

Fig. (2.2). Evolution of WT size and the power electronics seen from 1980 to 2020 (estimated). Orange circle: means the power coverage by power electronics, D: means diameter of the rotor [1].

Furthermore, to the fast progress in the full connected capacity, the size of single WT is also cumulative intensely to get a cheap price per generated kilowatt hour. The increasing tendencies of developing turbine dimension among 1980 and 2018 are shown in Fig. (**2.2**), where the development of PEs in the WTS is also shown.

It is well-known that the cutting-edge 8-MW WTs based on a diameter of 164 m have before now shown up in 2012 [5]. At present greatest of the turbine producers are developing products in the power range 4.5–8 MW, and it is estimated that gradually great WTs even up to 10-MW will seem in 2018, will be existent in the following years [1].

2. WIND TURBINE CONCEPTS

The general structure of Wind Energy Conversion System (WECS) containing an aerodynamic and electro-mechanical mechanism which transforms wind kinetic energy to electrical energy as displayed in Fig. (**2.3**).

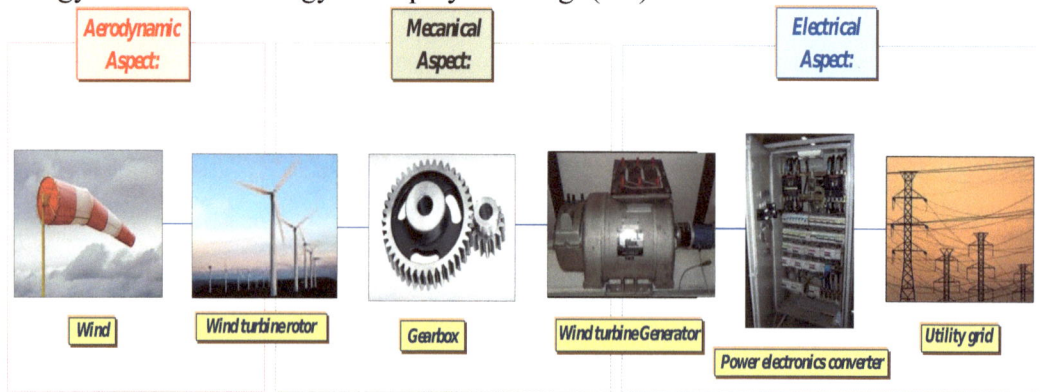

Aerodynamic Aspect: Mecanical Aspect: Electrical Aspect:

Wind Wind turbine rotor Gearbox Wind turbine Generator Power electronics converter Utility grid

Fig. (2.3). Power conversion stages in a typical WTS.

Wind power generation (WPG) uses whichever variable or fixed speed turbines which can be characterized into four (04) main types. The principal difference between these wind turbines categories is the method which is used in order to make imperfect the aerodynamic efficiency of the rotor regardless of numerous wind speed conditions. So these 04 (four) categories are illustrated below [6]:

2.1. Fixed Speed Wind Turbines (WT Type A)

Type "A" wind turbine (WT) is realized when an asynchronous squirrel-cage induction generator (SCIG) is associated directly to the network through a transformer. This type of WT desires an adjustment to prevent driving process during low wind speeds, and the consumption of reactive power present the main problem because there is no reactive power controller.

2.2. Partial Variable Speed Wind Turbine (VS-WT) Using Variable Rotor Resistance (Type B)

This type deals on a wound rotor induction generator (WRIG) be straight linked to the network. The measured resistances are associated in chain (series) with the rotor phase windings. In this manner, the global rotor resistances can be controlled, and consequently the output power and the slip can also be controlled.

2.3. VS-WT Using Partial Scale Power Converter (WT Type C)

This procedure, acknowledged as the doubly-fed induction generator (DFIG) notion, uses a variable speed regulator WT. The DFIG' stator phase windings are straight associated to the network, whereas the rotor is linked to a back-to-back converter *via* slip rings.

A-Advantages of the DFIG [7]

- The capacity of decoupling the reactive and active power by regulating the rotor terminal voltages.
- The DFIG is e in manufacture and economical than a permanent magnet synchronous generator (PMSG).
- Need a rigid network.
- The power converters is characteristically rated ±30% of the rated power, and this feature offers several advantages, for example, decreased converter cost, minimized filter cost and volume, fewer switching sufferers & harmonic injections into the network, and enhanced global efficiency [6].

B-Drawbacks of the DFIG [8 - 11]

- Requires slip-rings and gearbox, which will need regular upkeep.
- Limitation for fault ride through (FRT) ability and requires protection schemes.
- Reactive power capability limitation and complex control schemes.

2.4. VS-WT using Full Scale Power Converter (Type D)

This topology frequently uses a PMSG. The stator windings are associated to the network through a complete-scale power converter. This kind of WT adopts a gearless notion, instead of linking a gearbox to the generator, a driven generator is placed without a gearbox.

The drawbacks of the main two categories of WTs are:

1. Don't support any speed regulation,
2. Absence of reactive power compensation,
3. Need a rigid network,
4. Mechanical configuration must be capable to support great mechanical pressure produced by wind squalls.

Currently, DFIGs are most usually used in the WT manufacturing. In view of these advantages of the DFIG-based wind turbine-generator systems (WT-GS), this eBook will only concentrate on DFIGs and obviously in the next chapters; we will propose some detailed work about the mathematical models and control schemes.

3. CONTROL STRUCTURE OF WTS

The control of WT includes mutually fast and slow control dynamics [1] and [12, 13], as illustrated in Fig. (**2.5**), where an overall control structure for a WTS, counting generator, turbine, converter and filter. The WT concept can both be the type shown in Fig. (**2.4a and b**). Usually, the power flow has to be managed prudently in or out side of the generation system. The generated power should be controlled using mechanical parts (*e.g.,* pitch angle of blades). The full control system has to track the power production controls given by transmission system operator / distribution system operator. In the case of operation under grid fault, numerous subsystems in the WT which based on coordinated control such as, braking crowbar/chopper, grid/rotor side converters and pitch angle regulator are needed.

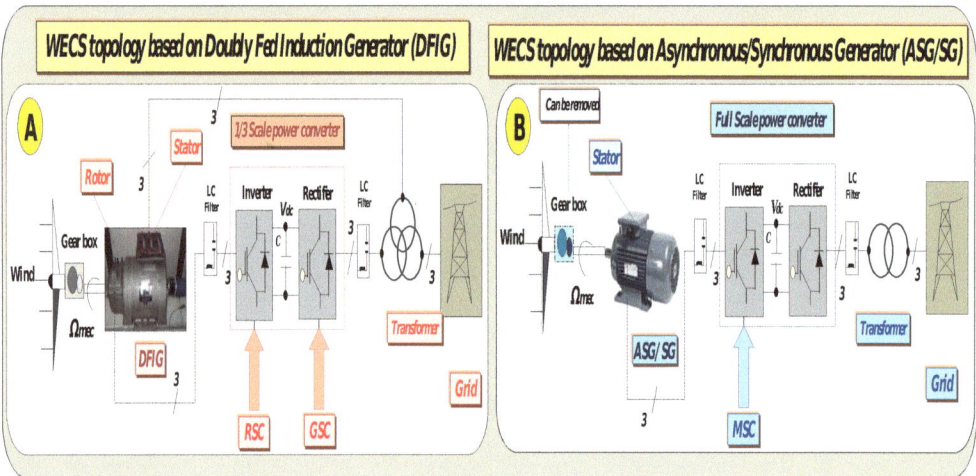

Fig. (2.4). A: Variable-speed wind turbine with partial-scale power converter and a DFIG and **B:** Variable-speed wind turbine with full-scale power converter.

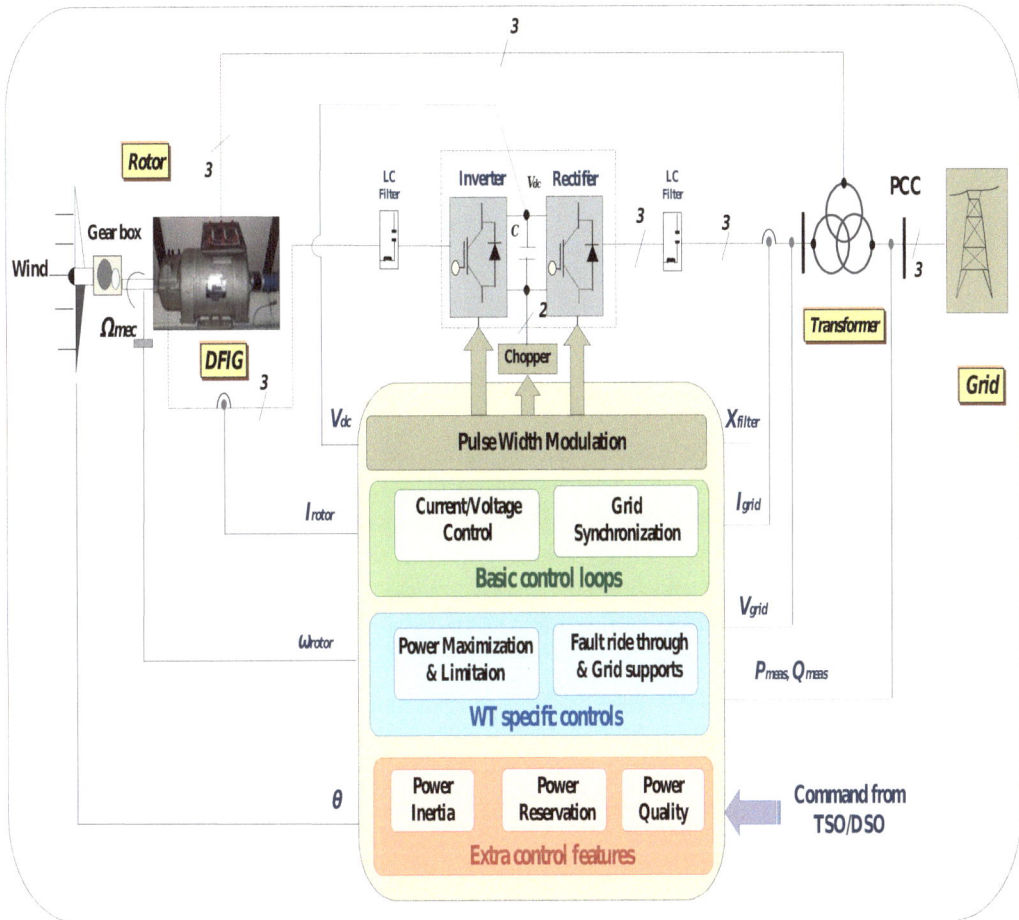

Fig. (2.5). Control structure for power electronics converter in WTS. (v_{dc}: dc-link voltage, I_{rotor}: rotor current, ω_{rotor}: rotational speed of rotor, θ: pitch angle of rotor blade, X_{filter}: filter impedance, I_{grid}: grid current, V_{grid}: grid voltage, P_{meas}: measured active power, Q_{meas}: Measured reactive power, PCC: point of common coupling).

Lastly, the elementary controls such as the grid synchronization, current regulation and dc bus balance have to be rapidly performed by the wind power converter [14 - 16].

4. LITERATURE SURVEY

In this part, a complete literature review describing DFIG based WT-GS will be illustrated. Further precisely, the related preceding studies and researches on the state of the art converter topologies applied in DFIG-based WT-GS, the modelling, the control approaches will be explained [17].

4.1. Modelling of a WTGS

The modelling of a WT-GS includes the power converter, the aerodynamic, the DFIG and the drive train system, refer to (Fig. **2.3**). Thus, this section of the study will lone focus on the DFIG modelling.

A- Modelling of DFIG

The doubly-fed induction machines (DFIMs) can be divided into four (04) categories. These categories are: the cascaded DFIM, the standard DFIM, the brushless DFIM and the single-frame cascaded DFIM [18]. However, only the standard and brushless types of DFIM have been used in WT-GS. In [19], the authors established the brushless-DFIG by using dual cascaded induction machines to remove the brushes, and applied a closed-loop SFO control system to reach reactive and active power control. In [20] and [21], in order to simplify the controller design, the authors implemented the synchronously reference frame for the reason that all the voltages and currents expressed below this frame will be of a DC nature. The DFIG model can generally be represented by reduced order models, which can produce a 3^{rd}order model by avoiding the derivative terms (DTs) of the Φ_s and first order model by ignoring together the DTs of the Φ_s and Φ_r [22]. The novel DFIG reduced 3rd order model is presented in [23], by neglecting the Rs and Ls through applying the *Laplace transformation*.

4.2. Control Strategies for a WT-GS

The control schemes for a WT-GS comprise MPPT control, the pitch angle control, and the DFIG control. The conventional and advanced control techniques for WT-GS are studied in details in this part.

A-Maximum Power Point Tracking (MPPT) Control

In order to reach the MPPT control, some controller schemes have been illustrated. The MPPT control can be principally divided into both types. They are the conventional and intelligent control schemes respectively. The classical control schemes can also be alienated into speed mode control and current mode control, which based on the tuning of reference values, which are the electromagnetic torque and active power (T_{em} & P_s) for current mode control [24, 25], and the rotating speed for the speed control [26, 27]. In [28], a comparative study between these two control approaches which determined that the current mode control has slow response with easy structure, whereas the speed mode control has fast response with difficult structure. The limits and discussions of these two control schemes were proposed in [29]. In reality, the wind speeds in overhead classical control schemes require to be precisely measured. On the other hand, the anemometer cannot exactly measure the wind speed for the reason that of complex landscape, the flow misrepresentation and tower shadow influence [30]. Therefore, some researches on maximum wind energy tracking devoid of wind speed measurement had been established in [31 - 33].

A.1 Intelligent Control

The intelligent control approaches generally used the fuzzy logic control (FLC) to the MPPT control. However, because the step disturbance is fixed; this control technique is frequently slow in speed. *e.g.,* a technique of using variable-step wind energy perturbation scheme to command the captured wind power was analysed in [34]. FLC based MPPT approaches have the benefits of having higher dynamic and stable performances, having robust speed control contrary to wind gusts and turbine undulations torque [35, 36].

A.2 Other Control Strategies

A new adaptive MPPT control was proposed in [37]. Knowing that, the wind speed and the maximum efficiency were estimated respectively by the output power generator and the maximum TSR (tip-speed ratio) tracker. In [38], a novel MPPT approach was proposed, in which there was no need for the information of wind turbine measurements and characteristic of the wind speed.

B-DFIG Control

Due to the large speed range operation (sub-synchronous and super-synchronous operations speed), the control of the DFIGs is more difficult than the control of a SCIG. In the last decades, several scientists have presented numerous types of DFIG control approaches, such as FOC, DTC/DPC, sensorless control, predictive control, and nonlinear control.

B.1 Field Oriented Control

FOC (Field oriented control) is generally used in DFIG controls due to two reasons: firstly, its capacity of regulating the motor speed more robustly, and secondly the low cost to make an FOC system. FOC moreover offers the aptitude to control independently the DFIG reactive and active powers [37].

B.2 Direct Torque/Power Control (DTC/DPC)

A novel direct power control (DPC) strategy is proposed in [39]. The aim of this approach is selecting suitable voltage vectors of the rotor side converter (RSC) in order to control the stator reactive and active powers of the DFIG. In order to mitigate or eliminate the problems or difficulties related with rotor flux estimation, the proposed control use only the estimated stator flux. The simulation results proved the robustness and effectiveness of the proposed control strategy especially under parameter changement of active and reactive powers, DC link voltage and rotor speed. In [40], the authors presented comparative simulation study between the Field Oriented Control (FOC) and Direct Power Control (DPC) for a 3kW DFIG is discussed, and in order to illustrate the feasibility of the proposed strategy an experimental test bench is made and tested under differents parameter condition. In [41], the authors focused on the analysis on the control of DFIG based high power-WT in the presence of voltage dips. The principal goal of the proposed control strategy is to remove or mitigate the need of the crowbar protection while Low-Depth Voltage Dips (LDVD) happens. Simulation results display the proposed control strategy that moderates the requirement of the crowbar protection through voltage dips.

B.3 Adaptive Nonlinear Control (MRAS Observer/MRAC Controller)

An adaptive controller for WT-DFIM was proposed by [42]. The controller behavior was tested with accurate electromagnetic DC/Power systems. Furthermore, the controller performances are tested in the existence of parameter changement and the voltage dips. In [43] proposed a sensorless control strategy for DFIGs in VS-WTS. In order to estimate the rotor position and rotor speed, the authors propose an extended Kalman filter (EKF). The control performance and estimation of the proposed sensorless method are demonstrated by simulation results. The performances of a model reference adaptive system (MRAS) observer and the EKF are compared for time-varying wind speeds. In [44] the authors proposed design and the experimental validation of the active and reactive power tuning based on model reference adaptive control (MRAC) of a WT-DFIG grid-connection. By means of the MPPT control, this regulation is realized under the synchronous speed. The experimental test bench is developed using 1KW DFIG with wind turbine emulator interfaced by dSPACE1104 card. The experimental results illustrate a good maximum wind power tracking despite of wind speed variation. Additionally the proposed controller displays a smooth tuning of the stator reactive and active power quantities switched between the generator and the electrical network.

B.4 Adaptive Disturbance Rejection Control (ADRC)

In [45], the authors proposed a new control to extract maximum electrical power from an aerodynamic power. In this new method an advanced order sliding mode supervisor is planned as an observer to concept the extractable power based on the condition that it operates. The simulations results display the important enhancement in performance of the backstepping nonlinear regulator exploiting this approach.

B.5 Sliding Mode Control (SMC)

In [46] a novel second or higher order sliding-mode control for a WT-DFIG was proposed. The objective of grid synchronization and power control are assumed by double diverse algorithms, calculated to control the RSC at a stable switching frequency. A process is moreover delivered that promises bump fewer transfer among the both controllers at DFIG grid-connection instant. In [47], the main control problem is the optimal power points of WT under wind-speed profiles and following them by means of adaptive online technique. In this paper, the disturbance are defined by wind speed and aerodynamics torque. Simulation results below diverse operational situations display the higher performance of the proposed control technique.

B.6 Backstepping Control (BSC)

In [48], the authors proposed a sort of combination between direct power control (DPC) and nonlinear backstepping-based algorithm of wind turbine DFIG below normal and especially harmonic grid-voltage. The simulation study among backstepping direct power control BS-DPC under normal grid voltage, look-up table DPC (LUT-DPC) and vector control (VC) through resonant controller, and confirm that the proposed BS-DPC comprehends the perfect decoupling control of reactive and active DFIG power, with superior dynamic performance than others. A BS-DPC approach below both unbalanced and balanced grid conditions for DFIG was proposed in [49]. Additionally, the enhanced strategy is proposed in order to attain diverse control goals under unstable grid situation without the requirement of decomposition of negative and positive sequence mechanisms. In [50], in order to regulate the reactive and active powers of WT, the decoupling control strategy has been applied. In order to control the WT power injected into the electrical network a proposed nonlinear Backstepping approach control has been proposed with integral actions. Lyapunov approach is applied in order to avoid the system nonlinearity by control laws derivation. The obtained simulation results verify the efficiency of the control schemes especially in: robustness, decoupling and dynamic response performances for diverse situations. In [51], the principal aim of the study is to derive a control strategy which will be applied to a bidirectional converter supplying a DFIG based on the combination of vector and backstepping control. Vector control approach has been initially carried out that get a decoupled reactive and active power with proportional-integral compensator (PI). In order to enhance some of the drawbacks of the PI compensator in terms of robustness, transient response and steady state error a backstepping approach has been used. Simulation study carried out on the DFIG generating both real powers extracted from the turbine and the required reactive power have shown good performances compared to those obtained by using PI compensator for the overall expected performance parameters.

B.7 Predictive Direct Power Control (PDPC) and Deadbeat Control

In [52]; the authors proposed a WT-DFIG control scheme *via* a PNC back to back three-level converter (3L-B2B) that maintains a unity power factor (PF=1) injected to the network and by means of a MPPT process in order to maximize the energy harvested from wind. A PDPC scheme regulates the GSC[1], by keeping the DC bus voltage (V_{DC}) at reference value. Whereas, a predictive direct torque control (PDTC) strategy controls the RSC machine, to control the power removal, the PF and DC bus balancing capacitors. Simulation results prove the efficiency of the advanced control techniques, counting the pitch angle control. In [53], authors proposed a new MPDPC strategy of DFIGs below unstable network voltage situations. The appropriate voltage vector is choose giving to an optimization cost function; hence the calculated reactive and active powers are controlled straight without the need of transformation, PWM[2] modulators or PI regulators and switching table. In [54], in order to regulate the frequency and voltage of a stand alone WGS[3], the application of the MPC[4] approach was proposed. To reduce computational work and to decrease mathematical difficulties, particularly in wide prediction horizon, an exponentially weighted functional model predictive control (FMPC) is tested. The planned regulator has been verified under load impedance and wind speed step variations. Simulation results presented high Wind-turbine performances in transient and steady states. In [55], the authors proposed a power control of DFIG for VS-WPG using a deadbeat control loop. The proposed process computes at each period the voltage vector to be provided to the rotor in order to assurance that the reactive and active power achieve their anticipated reference values. Simulations results prove the feasibility of the proposed control. In [56], the main aim is to offer the scheming and the demonstrating of DFIG-deadbeat power control. Thus, the deadbeat power controls goals the stator reactive and active power control using the discretized DFIG equations in stator flux orientation and synchronous coordinate system. Experimental results are offered in order to confirm the validation of the proposed regulator.

B.8 Input/Output Linearizing and Decoupling Control

In [57], the authors proposed a control scheme based on the input-output linearizing and decoupling control (I/OLDC) approach. An experimental test-bench based on 160 kW WT-DFIG, and the experimental results confirm that the I/OLDC offers enhanced decoupled DFIG-power control and dynamic responses. In [58], a novel type regulator approach for MPPT of DFIG has been proposed. The strategy adopts response of feedback linearization control (FLC) which based on Proportional-integral (PI) controllers in order to regulate the current and power DFIG loops respectively. The FLC can completely linearize and decouple the nonlinear system and offer perfect control performance in a large range operational situations. Nevertheless, the FLC approach might need some inaccessible state variables of the motorized scheme. Simulation results illustrate that the proposed strategy performs superior than the classical vector control (VC) scheme below changing process situations.

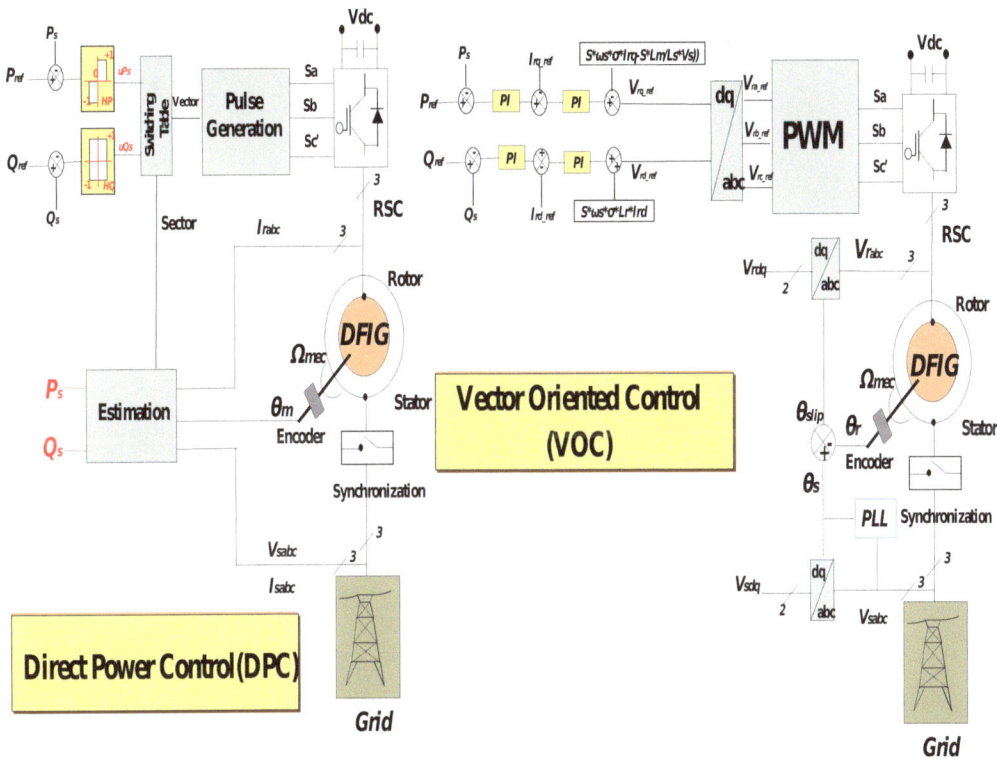

Fig. 2.6 cont.....

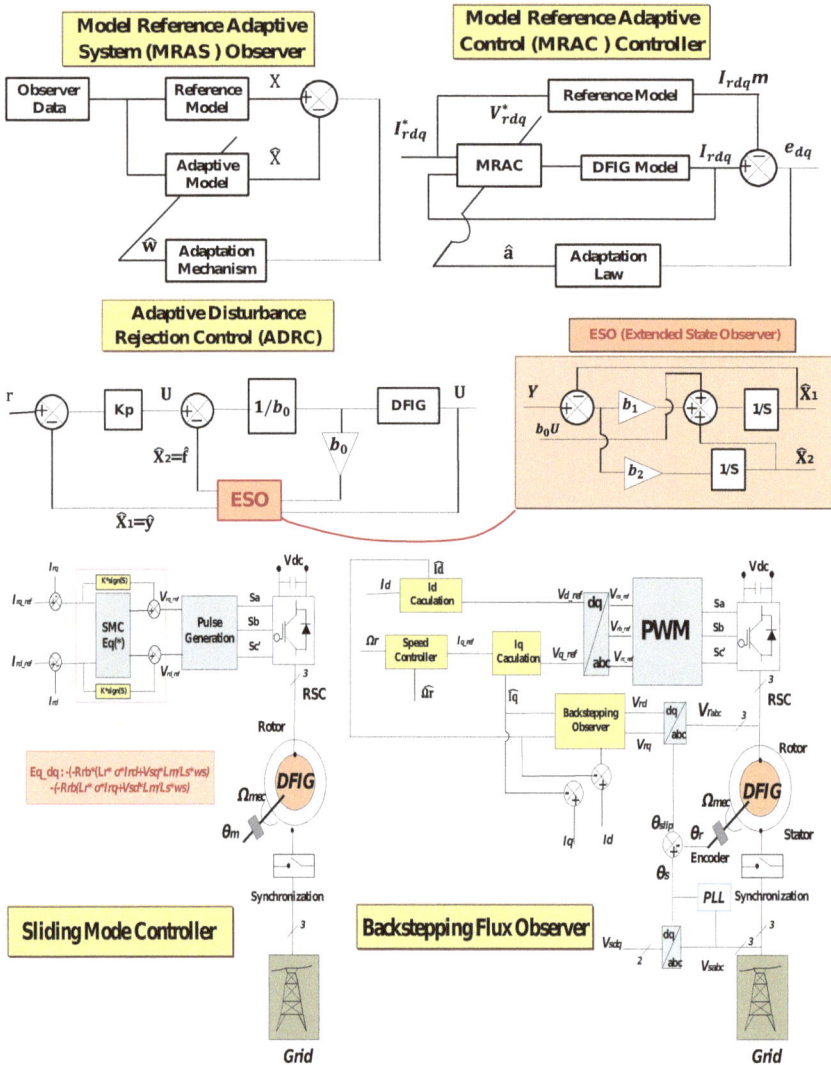

Fig. (2.6). Different control strategies of variable speed WT-DFIG for grid-connection.

NOTES
[1]Knowing that : GSC= Grid Side Converter & RSC=Rotor Side Converter.
[2]PWM=Pulse Width Modulation.
[3]WGS=Wind Generation System.
[4]MPC=Model Predictive Control.

REFERENCES

[1]   F. Blaabjerg, and K. Ma, "Future on power electronics for wind turbine systems", *IEEE J. Emerg. Sel. Top. Power Electron.,* vol. 1, no. 3, pp. 139-152, 2013.

[http://dx.doi.org/10.1109/JESTPE.2013.2275978]

[2]   REN21—Renewables 2012 Global Status Report, http://www.ren21.net

[3]   F. Blaabjerg, and K. Ma, "Wind Energy Systems", *Proc. of the IEEE,* vol. 105, 2017 pp. 2116-2131

[4]   G. Energy, Green Energy—The Road to a Danish Energy System without Fossil Fuels, http://www.klimakommissionen.dk/en-US/

[5]   Vestas Wind Power, Aarhus, Denmark, Wind Turbines Overview, 2011, www.vestas.com/.

[6]   H. S. Kim, and D. D.-C. Lu, "Wind Energy Conversion System from Electrical Perspective - A Survey", *Smart Grid and Renewable Energy,* vol. 1, no. 3, pp. 119-131, 2010.
[http://dx.doi.org/10.4236/sgre.2010.13017]

[7]   S. Muller, and M. Deicke, "Doubly fed induction generator systems for wind turbines", *IEEE. Ind. Appl. Magazine,* vol. 8, no. 3, pp. 26-33, 2002.

[8]   Wind Energy Background, Available: http://www.dolcera.com.

[9]   S. Engelhardt, I. Erlich, C. Feltes, J. Kretschmann, and F. Shewarega, "Reactive Power Capability of Wind Turbines Based on Doubly Fed Induction Generator", *IEEE Trans. Energy Conv.,* vol. 26, no. 1, pp. 364-372, 2011.
[http://dx.doi.org/10.1109/TEC.2010.2081365]

[10]   Asynchronous Generators, Available: http://www.windturbines.net

[11]   F. Raymond, W. Flumerfelt, and S.W. Su, "Wind turbines", *Access Science,* ©*Mc Graw-Hill Companies,* 2009. http://www.accessscience.com

[12]   "B. B.-Jensen, and M. H. Abdel-rahman, "Novel STATCOM controller for mitigating SSR and damping power system oscillations in a series compensated wind park", *IEEE Trans. Power Electron.,* vol. 25, no. 2, pp. 429-441, 2010.
[http://dx.doi.org/10.1109/TPEL.2009.2026650]

[13]   P. Rodriguez, A.V. Timbus, R. Teodorescu, M. Liserre, and F. Blaabjerg, "Flexible active power control of distributed power generation systems during grid faults", *IEEE Trans. Ind. Electron.,* vol. 54, no. 5, pp. 2583-2592, 2007.
[http://dx.doi.org/10.1109/TIE.2007.899914]

[14]   A. Timbus, M. Liserre, R. Teodorescu, P. Rodriguez, and F. Blaabjerg, "Evaluation of current controllers for distributed power generation systems", *IEEE Trans. Power Electron.,* vol. 24, no. 3, pp. 654-664, 2009.
[http://dx.doi.org/10.1109/TPEL.2009.2012527]

[15]   R. Teodorescu, M. Liserre, and P. Rodriguez, *Grid Converters for Photovoltaic and Wind Power Systems.* Wiley: New York, NY, USA, 2011.
[http://dx.doi.org/10.1002/9780470667057]

[16]   F. Blaabjerg, R. Teodorescu, M. Liserre, and A.V. Timbus, "Overview of control and grid synchronization for distributed power generation systems", *IEEE Trans. Ind. Electron.,* vol. 53, no. 5, pp. 1398-1409, 2006.
[http://dx.doi.org/10.1109/TIE.2006.881997]

[17]   S.K. Senapati, *"Modelling and Simulation of a Grid Connected Doubly Fed Induction Generator for Wind Energy Conversion System",* Master Thesis (English language), National Institute of Technology, Rourkela, INDIA, 2014.

[18]   A. Petersson, *"Analysis, Modeling and Control of Doubly-Fed Induction Generators for Wind Turbines",* Chalmers University of Technology, Ph.D. Thesis (English language), Sweden, 2005.

[19]   K. Protsenko, and D. Xu, "Dynamic Modeling of Doubly Fed Induction Generator Wind Turbines", *IEEE Trans. on Power Systems,* vol. 18, no. 2, pp. 803-809, 2003.
[http://dx.doi.org/10.1109/TPEL.2008.921187]

[20]   J.B. Ekanayake, L. Holdsworth, X. Wu, and N. Jenkins, "Dynamic Modeling of Doubly Fed Induction Generator Wind Turbines", *IEEE Trans. Power Syst.,* vol. 18, no. 2, pp. 803-809, 2003.
[http://dx.doi.org/10.1109/TPWRS.2003.811178]

[21]   I. Erlich, J. Kretschmann, J. Fortmann, S. Mueller-Engelhardt, and H. Wrede, "Modeling of Wind Turbine Based on Doubly-Fed Induction Generators for Power System Stability Studies", *IEEE Trans. Power Syst.,* vol. 22, no. 3, pp. 909-919, 2007.
[http://dx.doi.org/10.1109/TPWRS.2007.901607]

[22]   I. Erlich, and F. Shewarega, *Modeling of Wind Turbines Equipped with Doubly-Fed Induction Machines for Power System Stability Studies.* IEEE Power Systems Conference and Exposition, 2006, pp. 978-985.
[http://dx.doi.org/10.1109/PSCE.2006.296445]

[23]   A. Luna, F. Kleber de A, D. Lima, P. Santos, E. Rodríguez, H. Watanabe, and S. Arnaltes, "Simplified Modeling of a DFIG for Transient Studies in Wind Power Applications", *IEEE Trans. Ind. Electron.,* vol. 58, no. 1, pp. 9-20, 2011.
[http://dx.doi.org/10.1109/TIE.2010.2044131]

[24]   R. Pena, J.C. Clare, and G. Asher, "Doubly fed induction generator using back-to-back PWM converters and its application to variable-speed wind energy generation", *IEE Proc., Electr. Power Appl.,* vol. 143, no. 3, pp. 231-241, 1996.
[http://dx.doi.org/10.1049/ip-epa:19960288]

[25]   X. Zhen, Z. Xing, Y. Shuying, L. Qin, and Z. Wenfeng, "Study on Control Strategy of Maximum Power Capture For DFIG in Wind Turbine System", *2nd IEEE International Symposium on Power Electronics Distributed Generation Systems,* 2010pp. 110-115
[http://dx.doi.org/10.1109/PEDG.2010.5545875]

[26]   Y. Xiao, and P. Jia, "VSCF Wind Turbine Control Strategy for Maximum Power Generation", *Proc. of the 8th World Congress on Intelligent Control and Automation,* 2010 Jinan, China pp. 4781-4786.
[http://dx.doi.org/10.1109/WCICA.2010.5554559]

[27]   R. Datta, and V.T. Ranganathan, "A Method of Tracking the Peak Power Points for a Variable Speed Wind Energy Conversion System", *IEEE Trans. Energ. Convers.,* vol. 18, no. 1, pp. 163-168, 2003.
[http://dx.doi.org/10.1109/TEC.2002.808346]

[28]   B. Shen, B. Mwinyiwiwa, Y. Zhang, and B-T. Ooi, "Sensorless Maximum Power Point Tracking of Wind by DFIG Using Rotor Position Phase Lock Loop (PLL)", *IEEE Trans. Power Electron.,* vol. 24, no. 4, pp. 942-951, 2009.
[http://dx.doi.org/10.1109/TPEL.2008.2009938]

[29]   C. Shao, X. Chen, and Z. Liang, "Application Research of Maximum Wind energy Tracing Controller Based Adaptive Control Strategy in WECS", *IEEE 5th International Power Electronics and Motion Control Conf.,* 2006 pp. 1-5.

[30]   E. Koutroulis, and K. Kalaitzakis, "Design of a Maximum Power Tracking System for Wind-Energy-Conversion Applications", *IEEE Trans. on Indust. Electron.,* vol. 53, no. 2, pp. 486-494, 2006.
[http://dx.doi.org/10.1109/TIE.2006.870658]

[31]   W. Qiao, W. Zhou, J. M. Aller, and R. G. Harley, "Wind Speed Estimation Based Sensorless Output Maximization Control for a Wind Turbine Driving a DFIG", *IEEE Trans. Power Electron.,* vol. 23, no. 3, pp. 1156-1169, 2008.

[32]   S. Chondrogiannis, and M. Barnes, "Stability of Doubly-Fed Induction Generator under Stator Voltage Oriented Vector Control", *IET Renew. Power Gener.,* vol. 2, no. 3, pp. 170-180, 2008.
[http://dx.doi.org/10.1049/iet-rpg:20070086]

[33]   C. Batlle, "A. D.-Cerezo and R. Ortega, "A Stator Voltage Oriented PI Controller For The Doubly-Fed Induction Machine", *Proc. of the American Control Conference,* 2007 New York City, USA, pp. 5438-5443.

[34]   C. Wang, and G. Weiss, "Integral Input-to-State Stability of the Drive- Train of a Wind Turbine", *Proc. of the 46th IEEE Conference on Decision and Control,* 2007 New Orleans, LA, USA, pp. 6100-6105.

[35]   Y. Lei, A. Mullane, G. Lightbody, and R. Yacamini, "Modeling of the Wind Turbine With a Doubly Fed Induction Generator for Grid Integration Studies", *IEEE Trans. Energ. Convers.,* vol. 21, no. 1, pp. 257-264, 2006.
[http://dx.doi.org/10.1109/TEC.2005.847958]

[36]   M. Tazil, V. Kumar, R.C. Bansal, S. Kong, Z.Y. Dong, W. Freitas, and H.D. Mathur, "Three-phase doubly fed induction generators: an overview", *IET Electr. Power Appl.,* vol. 4, no. 2, pp. 75-89, 2010.
[http://dx.doi.org/10.1049/iet-epa.2009.0071]

[37]   D. Casadei, F. Profumo, G. Serra, and A. Tani, "FOC and DTC: Two Viable Schemes for Induction Motors Torque Control", *IEEE Trans. Power Electron.,* vol. 17, no. 5, pp. 779-787, 2002.
[http://dx.doi.org/10.1109/TPEL.2002.802183]

[38]   K.C. Wong, S.L. Ho, and K.W.E. Cheng, "Direct Torque Control of a Doubly fed Induction Generator with Space Vector Modulation", *Electr. Power Compon. Syst.,* vol. 36, no. 12, pp. 1337-1350, 2008.
[http://dx.doi.org/10.1080/15325000802258331]

[39]   L. Xu, and P. Cartwright, "Direct Active and Reactive Power Control of DFIG for Wind Energy Generation", *IEEE Trans. Energy Conv.,* vol. 21, no. 3, 2006.
[http://dx.doi.org/10.1109/TEC.2006.875472]

[40]   S.-T. Jou, S.-B. Lee, Y.-B. Park, and K.-B. Lee, "Direct Power Control of a DFIG in Wind Turbines to Improve Dynamic Responses", *JPE J. Power. Electron.,* vol. 9, no. 5, pp. 781-790, 2012.

[41]   Y.S. Rao, and A.J. Laxmi, "Direct Torque Control of Doubly Fed Induction Generator based Wind Turbine under Voltage Dips", *Int. J. Adv. Eng. Technol.,* vol. 3, no. 2, pp. 711-720, 2012.

[42]   J. M. Mauricio, and A. E. Leon, "An Adaptive Nonlinear Controller for DFIM-Based Wind Energy Conversion Systems", *IEEE Trans. Energ. Conv.,* vol. 23, no. 4, 2008.

[43]   M. Abdelrahem, C. Hackl, and R. Kennel, "Sensorless Control of Doubly-Fed Induction Generators in Variable-Speed Wind Turbine Systems", *International Conference on Clean Electrical Power (ICCEP), IEEE conference,* 2015 pp. 406-413
[http://dx.doi.org/10.1109/ICCEP.2015.7177656]

[44]   S. Abdeddaim, A. Betka, S. Drid, and M. Becherif, "Implementation of MRAC controller of a DFIG based variable speed grid connected wind turbine", *Energy Convers. Manage.,* vol. 79, pp. 281-288, 2014.
[http://dx.doi.org/10.1016/j.enconman.2013.12.003]

[45]   A. Tohidi, H. Hajieghrary, and M.A. Hsieh, "Adaptive Disturbance Rejection Control Scheme for DFIG-Based Wind Turbine: Theory and Experiments", *IEEE Trans. Ind. Appl.,* vol. 52, no. 03, pp. 2006-2015, 2016.
[http://dx.doi.org/10.1109/TIA.2016.2521354]

[46]   A. Susperregui, M.I. Martinez, G. Tapia, and I. Vechiu, "Second-Order Sliding-Mode Controller Design and Tuning for Grid Synchronization and Power Control of a Wind Turbine-Driven Doubly Fed Induction Generator", *IET Renew. Power Gener.,* vol. 7, no. 5, pp. 540-551, 2013.
[http://dx.doi.org/10.1049/iet-rpg.2012.0026]

[47]   A. Tohidi, A. Shamsaddinlou, and A.K. Sedigh, "Multivariable input-output linearization sliding mode control of DFIG based wind energy conversion system", *9th Asian Control Conference (ASCC), IEEE Conference,* 2013 pp. 1-6
[http://dx.doi.org/10.1109/ASCC.2013.6606347]

[48]   P. Xiong, and D. Sun, "Backstepping-Based DPC Strategy of a Wind Turbine-Driven DFIG under Normal and Harmonic Grid Voltage", *IEEE Trans. Power Electron.,* vol. 31, no. 6, pp. 4216-4225, 2016.

[http://dx.doi.org/10.1109/TPEL.2015.2477442]

[49]   X. Wang, D. Sun, and Z.Q. Zhu, "Resonant-Based Backstepping Direct Power Control Strategy for DFIG under Both Balanced and Unbalanced Grid Conditions", *IEEE Trans. Ind. Appl.,* vol. 53, no. 5, pp. 4821-4830, 2017.
[http://dx.doi.org/10.1109/TIA.2017.2700280]

[50]   M. Doumi, A. Ghani Aissaoui, A. Tahour, M. Abid, and K. Tahir, "Nonlinear Integral Backstepping Control of Wind Energy Conversion System Based on a Double-Fed Induction Generator", *Przegląd Elektrotechniczny,* vol. xx, no. xx, pp. 130-135, 2016.
[http://dx.doi.org/10.15199/48.2016.03.32]

[51]   R. Rouabhi, R. Abdessemed, A. Chouder, and A. Djerioui, "Hybrid Backstepping Control of a Doubly Fed Wind Energy Induction Generator", *The Mediterranean J Measurement and Control.,* vol. 11, no. 1, 2015.

[52]   J. Sayritupac, E. Albánez, J. Rengifo, J.M. Aller, and J. Restrepo, "Predictive control strategy for DFIG wind turbines with maximum power point tracking using multilevel converters", *IEEE Workshop on Power Electronics and Power Quality Applications (PEPQA),* 2015 pp. 1-6.
[http://dx.doi.org/10.1109/PEPQA.2015.7168207]

[53]   J. Hu, J. Zhu, and D.G. Dorrell, "Predictive Direct Power Control of Doubly Fed Induction Generators Under Unbalanced Grid Voltage Conditions for Power Quality Improvement", *IEEE Trans. Sustainable Energ.,* vol. 6, no. 3, pp. 943-950, 2015.
[http://dx.doi.org/10.1109/TSTE.2014.2341244]

[54]   A.M. Kassem, "Predictive Voltage Control of Stand Alone Wind Energy Conversion System", *WSEAS Trans. on Systems and Control,* vol. 7, no. 3, pp. 97-107, 2012.

[55]   A. J. Sguarezi Filho, and E. Ruppert, "A Deadbeat Active and Reactive Power Control for Doubly Fed Induction Generator", *Taylor & Francis Group, LLC, Electric Power Components and Systems,* vol. 38, pp. 592-602, 2010.
[http://dx.doi.org/10.1080/15325000903376966]

[56]   D.P. Bhattacharya, Ed., *A. J. Sguarezi filho and E. Ruppert, "Modeling and Designing a Deadbeat Power Control for Doubly-Fed Induction Generator", Chapter n°:06 from eBook: "Wind Energy Management.,* 2017, pp. 113-128.

[57]   L. Zhang, X. Cai, and J. Guo, "Simplified input-output linearizing and decoupling control of wind turbine driven doubly-fed induction generators", In: *IEEE 6ᵗʰ International Power electronics and motion Control Conference,* 2009.

[58]   X. Lin, K.S. Xiahou, Y. Liu, Y.B. Zhang, and Q.H. Wu, *"Maximum Power Point Tracking of DFIG-WT Using Feedback Linearization Control Based Current Regulators", IEEE Innovative Smart Grid Technologies - Asia.* ISGT-Asia, 2016, pp. 718-723.

Indirect Power Control (IDPC) of DFIG Using Classical & Adaptive Controllers Under MPPT Strategy

Abstract: In this chapter, we present a comparative study of conventional Indirect Power Control (IDPC) algorithm of DFIG-Wind turbine in grid-connection mode, using PI and PID controllers *via* Maximum power point tracking (MPPT) strategy. Firstly, the conventional IDPC based on PI controllers will be described using simplified model of DFIG through stator flux orientation and wind-turbine model. The MPPT strategy is developed using Matlab/Simulink® with two wind speed profiles in order to ensure the robustness of wind-system by maintaining the Power coefficient (Cp) at maximum value and reactive power at zero level; regardless unexpectedF wind speed variation. Secondly, the rotor side converter (RSC) and Grid side converter (GSC) are illustrated and developed using Space vector modulation (SVM) in order to minimize the stress and the harmonics and to have a fixed switching frequency. In this context, the switching frequency generated by IDPC to control the six IGBTs of the inverter (RSC), and this control algorithm works under both Sub- and Super-synchronous operation modes and depending to the wind speed profiles. The quadrants operation modes of the DFIG are described in details using real DFIG to show the power flow under both modes (motor and generator in the four (04) quadrants. Finally, the conventional IDPC have several drawbacks as: response time, power error and overshoot. In this context, the PID and MRAC (adaptive regulator) controllers are proposed instead of the PI to improve the wind-system performances *via* MPPT strategy with/without robustness tests. The obtained simulation results under Matlab/Simulink® show high performances (in terms of power error, power tracking and response time) in steady and transient states despite sudden wind speed variation, whereas big power error and remarkable overshoot are noted using robustness tests, so the proposed IDPC can not offer big improvement under parameter variation.

Keywords: Indirect Power Control (IDPC) algorithm, Maximum power point tracking (MPPT) strategy, Model Reference Adaptive Control (MRAC), Space vector modulation (SVM).

1. INTRODUCTION

Wind energy conversion system (WECSs) based on the doubly-fed induction generator (DFIG) dominated the wind power generations due to the outstanding advantages, including small converters rating around 30% of the generator rating,

Fayssal Amrane & Azeddine Chaiba

lower converter cost. Several novel control strategies have been investigated in order to improve the DFIG operation performance [1 - 4].

Nowadays, since DFIG-based WECSs (Fig. **3.1**) are mainly installed in remote and rural areas [2]. In literature [5] vector control is the most popular method used in the DFIG-based wind turbines (WTs). In most applications, the proportional-integral (PI) controller based Direct Power Vector Control (DPC) scheme is used to control DFIG in wind energy conversion systems [6, 7]. Although this control scheme is easy to implement and based on linear controller (as PI controller) which generates a fast and robust power response without the need of complex structure and algorithms. It has some drawbacks. One of the most important drawbacks of this control scheme is that the performance of the PVC (known also by Indirect Power Control 'IDPC') scheme largely depends on the tuning of the PI controller's parameters (K_p and K_i) and high power ripple during a steady state. Another drawback of this controller is that its performance also depends on the accuracy of the machine parameters and on grid voltage conditions such as harmonic level, distortion, *etc*. [7].

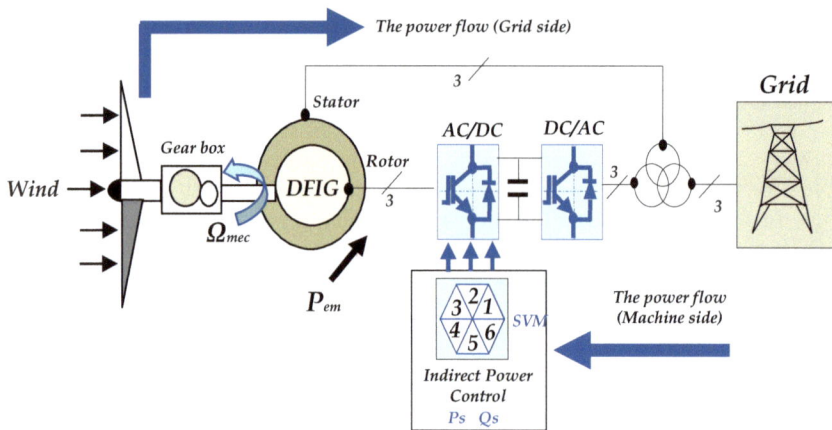

Fig. (3.1). Schematic diagram of wind-turbine DFIG based on conventional indirect power control (IDPC).

However, the switching frequency of the converter is still affected significantly by active and reactive power variations and operating speed. In this context, Space Vector modulation (SVM) will present the best solution [8, 9] compared to Pulse Width Modulation (PWM) in order to minimize the harmonic and keeping the fixed switching frequency of AC-DC-AC converters (especially in Rotor Side Converter 'RSC'). To extract the maximum power despite sudden variations in wind speed, the maximum power point tracking (MPPT) strategy [10 - 14] is proposed, the stator active power is extracted from wind power and stator reactive power is maintained at zero level to ensure unity power.

Model Reference Adaptive Control (MRAC) is a kind of control method that follows the response signal at the output of reference model. It has the advantages of simple structure and fast and stable reconfiguration. The general idea underpinning MRAC is to incorporate a reference model to acquire the preferred closed-loop reactions. MRAC has the ability to control a system that undergoes parameter and/or environmental variations. It designs the mechanism law and adjustments technique to drive the desired trajectories for the system to track the reference model output [15, 16]. The analysis and performance of several model reference adaptive system (MRAS) observers for sensorless vector control of DFIG is proposed in [17]. A model reference adaptive control (MRAC) speed estimator speed sensorless direct torque and flux control of an induction motor is proposed in [18] to achieve high performance sensorless drive. In order to improve the performance of the IDPC scheme, Proportional-Integral-Derivate (PID) [19] and Model reference adaptive control (MRAC) [20] are proposed instead the PI controllers [21 - 23] to hence wind-system performance in terms of power tracking, power error and overshoot.

In this chapter, a WECS's study was presented which described in details the mathematical model of; the DFIG, the IDPC, the turbine, the MPPT strategy, the GSC, the RSC based on SVM and finally the comparative simulation study of the conventional and proposed IDPC with/without robustness tests to show the improvement offered by the proposed control in transient and steady states.

2. MATHEMATICAL MODEL OF DFIG

The generator chosen for the conversion of wind energy is a double-fed induction generator, *DFIG* modeling described in the two-phase reference (*Park*). The general electrical state model of the induction machine obtained using Park transformation is given by the following equations [24, 25]:

Stator and rotor voltages:

$$V_{sd} = R_s * I_{sd} + \frac{d}{dt}\Phi_{sd} - \omega_s * \Phi_{sq}. \tag{3.1}$$

$$V_{sq} = R_s * I_{sq} + \frac{d}{dt}\Phi_{sq} + \omega_s * \Phi_{sd}. \tag{3.2}$$

$$V_{rd} = R_r * I_{rd} + \frac{d}{dt}\Phi_{rd} - (\omega\|s - \omega_r) * \Phi_{rq}. \tag{3.3}$$

$$V_{rq} = R_r * I_{rq} + \frac{d}{dt} \Phi_{rq} + \left(\omega \| s - \omega_r \right) * \Phi_{rd}. \tag{3.4}$$

Stator and rotor fluxes:

$$\Phi_{sd} = L_s * I_{sd} + L_m * I_{rd}. \tag{3.5}$$

$$\Phi_{sq} = L_s * I_{sq} + L_m * I_{rd}. \tag{3.6}$$

$$\Phi_{rd} = L_r * I_{rd} + L_m * I_{sd}. \tag{3.7}$$

$$\Phi_{rd} = L_r * I_{rd} + L_m * I_{sd}. \tag{3.8}$$

The electromagnetic torque is given by:

$$T_{em} = P * L_m * \left(I_{rd} * I_{sq} - I_{rq} * I_{sd} \right). \tag{3.9}$$

And its associated motion equations are:

$$T_{vis} = f * \Omega_{mec}. \tag{3.10}$$

$$T_{em} - T_r = \frac{J * d}{dt} \Omega_{mec} + T_{vis}. \tag{3.11}$$

$$J = \frac{J_{turbine}}{G^2} + J_g. \tag{3.12}$$

Where; R_s, R_r, L_r and L_s are respectively the resistances and the inductances of the stator and the rotor of the *DFIG*. V_{sd}, V_{sq}, V_{rd}, V_{rq}, I_{sd}, I_{sq}, I_{rd}, I_{rq}, Φ_{sd}, Φ_{sq}, Φ_{rd} and Φ_{rq} respectively represent the components along the *d and q* axes of the stator and rotor voltages, currents and flux. T_{em}, T_r, T_{vis}, T_{aero} and $T_{gearbox}$ present respectively the electromagnetic, load, viscous, aerodynamic and gearbox torques. J_g, $J_{Turbine}$ and J are the generator', turbine' and total inertia in *DFIG's* rotor respectively, Ω_{mec} is the mechanical speed, and G is the gain of gear box.

The active and reactive powers of DFIG's stator and rotor are respectively:

$$P_s = V_{sd} * I_{sd} + V_{sq} * I_{sq} \tag{3.13}$$

$$Q_s = V_{sq} * I_{sd} - V_{sd} * I_{sq} \tag{3.14}$$

$$P_r = V_{rd} * I_{rd} + V_{rq} * I_{rq} \tag{3.15}$$

$$Q_r = V_{rq} * I_{rd} - V_{rd} * I_{rq} \tag{3.16}$$

The frequency of the stator voltage being imposed by the electrical network (*grid*), the pulsation of the rotational currents (*slip*) is given by:

$$\omega_{slip} = \omega_s - \omega_r \wedge (\omega \| r = P * \Omega_{mec}) \tag{3.17}$$

Where: ω_r *and* ω_s represent respectively the pulsations of rotor and stator voltages in *rad/sec*. The angles θ_s *and* θ_s are obtained respectively by integration of ω_s *and* ω_s.

$$\theta_s = \int_0^t \omega_s . dt + \theta_{s0} . \tag{3.18}$$

$$\theta_r = \int_0^t \omega_r . dt + \theta_{r0} \tag{3.19}$$

3. CONVENTIONAL INDIRECT POWER CONTROL (*IDPC*) OF DFIG

The *DFIG* model can be described by the following state equations in the synchronous reference frame whose *axis "d"* is aligned with the stator flux vector as shown in Fig. (**3.2**) [24, 25]:

$$\Phi_{sd} = L_s * i_{sd} + L_m * i_{rd} \tag{3.20}$$

$$0 = L_s * i_{sq} + L_m * i_{rq} \tag{3.21}$$

From the equations of the direct and quadrature components of the stator flux, we obtain the following expressions of the stator currents:

$$I_{sq} = \frac{-L_m}{L_s} * I_{rq} . \tag{3.23}$$

$$i_{sd} = \frac{\Phi_{sd}}{L_s} - \frac{L_m}{L_s} * i_{rd} . \tag{3.22}$$

By replacing in the equations of the rotor flux, it finds:

$$\Phi_{rd} = L_r \left(1 - \frac{L_m^2}{L_s * L_r} \right) * I_{rd} + \frac{L_m}{L_s} * \Phi_{sd} = L_r * \sigma * I_{rd} + \frac{L_m}{L_s} * \Phi_{sd}. \tag{3.24}$$

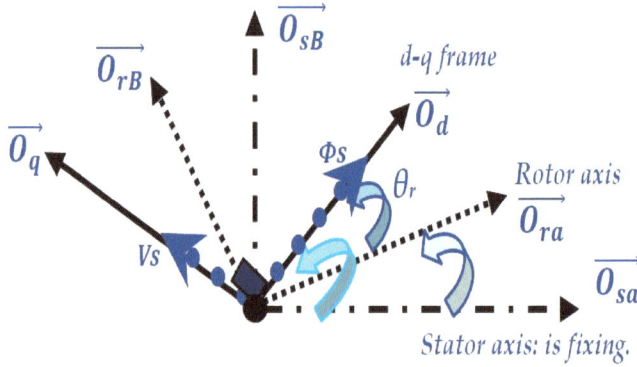

Fig. (3.2). Stator and rotor flux vectors in the synchronous d-q frame.

$$\Phi_{rq} = L_r \left(1 - \frac{L_m^2}{L_s * L_r} \right) * I_{rq} = L_r * \sigma * I_{rq}. \tag{3.25}$$

σ: is the dispersion coefficient between windings *d and q*:

$$\sigma = 1 - \frac{L_m^2}{L_s . L_r}. \tag{3.26}$$

By integrating the equations of the stator currents and the rotor flux into the equations of the stator and rotor voltages, we will have:

$$V_{sd} = \frac{R_r}{L_s} * \Phi_{sd} - \frac{R_r}{L_s} * L_m * I_{rd} + \frac{d\Phi_{sd}}{dt}. \tag{3.27}$$

$$V_{sq} = \frac{R_r}{L_s} * L_m * I_{rq} + \omega_s * \Phi_{sd}. \tag{3.28}$$

$$V_{rd} = R_r * I_{rd} + \frac{L_r * \sigma * dI_{rd}}{dt} + \frac{\frac{L_m}{L_s} * d\Phi_{sd}}{dt} - L_r * (\omega_s - \omega_r) * \sigma * I_{rq}. \tag{3.29}$$

$$V_{rq} = R_r * I_{rq} + \frac{L_r * \sigma * dI_{rq}}{dt} + \frac{L_m}{L_s} * (\omega_s - \omega_r) * \Phi_{sd} + L_r * (\omega_s - \omega_r) * \sigma * I_{rd}. \tag{3.30}$$

For control of the generator, expressions are established showing the relationship between the currents and the rotor voltages applied to it:

$$\frac{dI_{rd}}{dt} = \frac{1}{\sigma * L_r} * \left(V_{rd} - R_r * I_{rd} - e_d\right).$$

$$\tag{3.31}$$

$$\frac{dI_{rq}}{dt} = \frac{1}{\sigma * L_r} * \left(V_{rq} - R_r * i_{rq} - e_q - e_\Phi\right).$$

$$\tag{3.32}$$

With the electromotive forces:

$$e_d = \frac{\frac{L_m}{L_s} * d\Phi_{sd}}{dt} - L_m * \left(\omega_s - \omega_r\right) * \sigma * I_{rq}.$$

$$\tag{3.33}$$

$$e_\Phi = \frac{L_m}{L_s} * \left(\omega_s - \omega_r\right) * \Phi_{sd}.$$

$$\tag{3.34}$$

$$e_q = L_r * \left(\omega_s - \omega_r\right) * \sigma * I_{rd}.$$

$$\tag{3.35}$$

With the electromotive forces: On the other hand the expression of the electromagnetic torque becomes:

$$T_{em} = P * \left(\frac{L_m}{L_s}\right) * \Phi_{sd} * I_{rq}.$$

$$\tag{3.36}$$

3.1. Relationship Between Rotor Voltages and Rotor Currents (Generally Form)

If we neglect the stator resistance, which is a valid assumption for the medium and large power machines used in wind energy. So we simplify the expressions of the stator voltages as follows:

$$V_{sd} = \frac{d\Phi_{sd}}{dt}.$$

$$\tag{3.37}$$

$$V_{sq} = \omega_s * \Phi_{sd}.$$

$$\tag{3.38}$$

If we assume that the electrical network is stable, having for simple voltage V_s and for stator pulsation ω_s (*supposed constant*), this leads to a constant stator flux Φ_{sd}, with a zero variation in steady state, which makes it possible to write:

$$V_{sd} = 0, V_{sq} = \omega_s * \Phi_{sd}. \tag{3.39}$$

So the decoupling relationships are:

$$e_d^* = +L_r * \sigma * (\omega_s - \omega_r) * I_{rq}^*. \tag{3.40}$$

$$e_q^* = -L_r * \sigma * (\omega_s - \omega_r) * I_{rd}^*. \tag{3.41}$$

For the equation of the slip[1] (**S**), we replace ω_s-ω_r as follows:

$$S = \frac{\omega_s - \omega_r}{\omega_s} \text{ and } \omega_s - \omega_r = S * \omega_s \tag{3.42}$$

$$e_d^* = +L_r * \sigma * S * \omega_s * I_{rq}^*. \tag{3.43}$$

$$e_q^* = -L_r * \sigma * S * \omega_s * I_{rd}^*. \tag{3.44}$$

Regulators of rotor currents:

$$V_{rd}^* = PI (I_{rd}^* - I_{rd\_meas}). \tag{3.45}$$

$$V_{rq}^* = PI (I_{rq}^* - I_{rq\_meas}). \tag{3.46}$$

3.2. Relationship Between Stator Power and Rotor Currents

In a two-phase reference, the stator active and reactive powers of a *DFIG* are written as follows:

$$P_s = V_{sd} * I_{sd} + V_{sq} * I_{sq} \tag{3.47}$$

$$Q_s = V_{sq} * I_{sd} - V_{sd} * I_{sq} \tag{3.48}$$

With $V_{sd} = 0$ and replacing "I_{sd}" and "I_{sq}" with their expressions, with $\Phi_s = V_s / \omega_s$, we gets:

$$P_s^* = -V_s * \frac{L_m}{L_s} * I_{rq}^*. \tag{3.49}$$

$$Q_s^* = \frac{V_s^2}{\omega_s * L_s} - V_s * \frac{L_m}{L_s} * I_{rd}^*. \tag{3.50}$$

Regulators of stator powers:

$$I_{rd}^* = PI\ (Q_s^* - Q_{s\_meas}). \tag{3.51}$$

$$I_{rq}^* = PI\ (P_s^* - P_{s\_meas}). \tag{3.52}$$

3.3. Relationship Between Rotor Voltages and Rotor Currents (Detailed Form)

By replacing the stator currents by their expressions in the rotor flux expressions, we will have:

$$\Phi_{rd} = \left(L_r - \frac{L_m^2}{L_s} \right) * I_{rd} + \frac{L_m * V_s}{\omega_s * L_s}. \tag{3.53}$$

$$\Phi_{rq} = \left(L_r - \frac{L_m^2}{L_s} \right) * I_{rq}. \tag{3.54}$$

The expressions of the rotor fluxes of *d and q* axes are introduced into the expressions of the two-phase rotor voltages. We are getting:

$$V_{rd} = R_r * I_{rd} + \frac{\left(L_r - \frac{L_m^2}{L_s} \right) * dI_{rd}}{dt} - S * \omega_s * \left(L_r - \frac{L_m^2}{L_s} \right) * I_{rq} \tag{3.55}$$

$$V_{rq} = R_r * I_{rq} + \frac{\left(L_r - \frac{L_m^2}{L_s} \right) * dI_{rq}}{dt} + S * \omega_s * \left(L_r - \frac{L_m^2}{L_s} \right) * I_{rd} + \frac{S * L_m * V_s}{L_s}. \tag{3.56}$$

In steady state, the terms involving derivatives of out of phase rotor currents disappear, so we can write:

$$V_{rd} = R_r * I_{rd} - S * \omega_s * \left(L_r - \frac{L_m^2}{L_s} \right) * I_{rq}. \tag{3.57}$$

$$V_{rq} = R_r * I_{rq} + S * \omega_s * \left(L_r - \frac{L_m^2}{L_s} \right) * I_{rd} + \frac{S * L_m * V_s}{L_s}. \tag{3.58}$$

With: V_{rd}, V_{rq}: are the out of phase components of the rotor voltages to be imposed on the generator (*DFIG*) to obtain the desired rotor currents.

$\left(L_r - \dfrac{L_m^2}{L_s} \right)$:*Is the coupling term between the both axes (d and q);*

$\left(\dfrac{S * L_m * V_s}{L_s} \right)$: *Represents an electromotive force dependent on the speed rotation.*

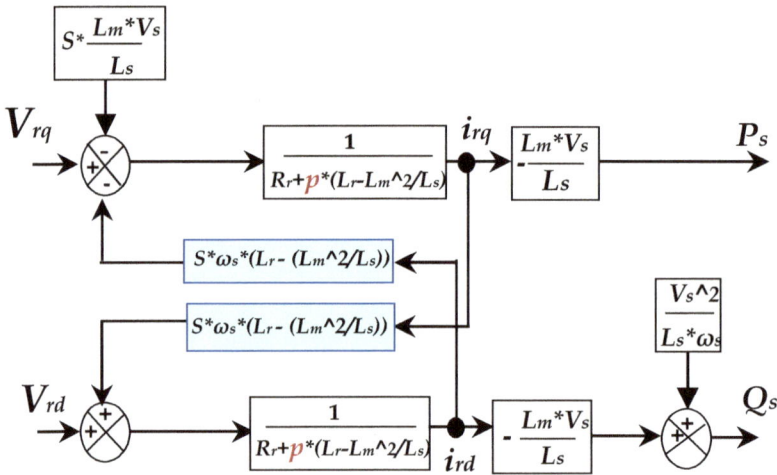

Fig. (3.3). The DFIG simplified power model.

Fig. (**3.3**) illustrates the simplfied model of the *DFIG* and Fig. (**3.4**) presents the conventional Indirect Power Control (*IDPC based on PI controllers*) of *DFIG*.

3.4. Synthesis of the Proportional-Integral (PI) Regulator

To carry out the looped control; conventional regulators *'PI'* are used. This type of regulator ensures a zero static error thanks to the integral action while the speed of the response is established by the proportional action. The reference quantities for these regulators will be active and reactive power [21 - 23]. Fig. (**3.5a&b**) show the *PI* controller structure with the closed loop transfer function of the wind-system based on *PI*.

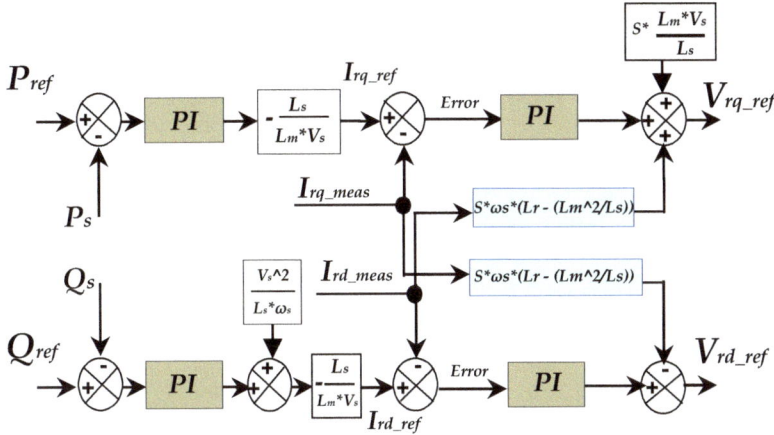

Fig. (3.4). DFIG's conventional indirect power control.

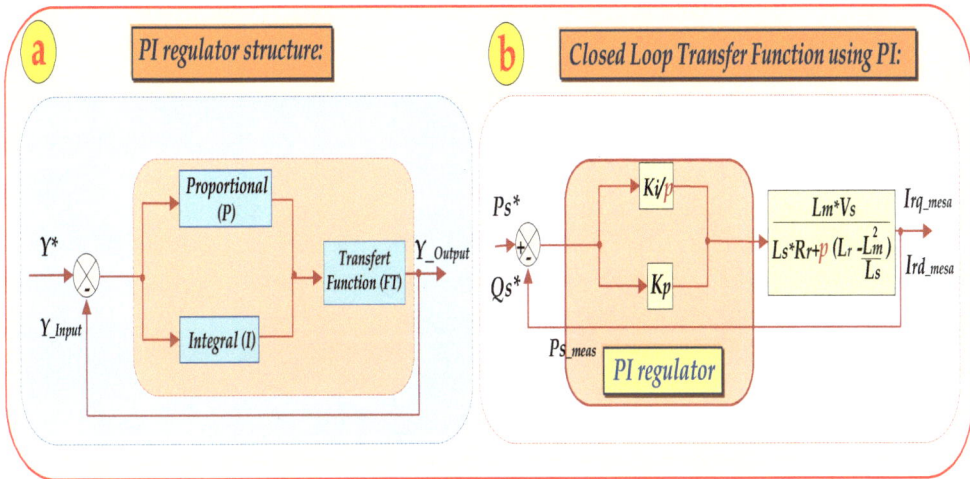

Fig. (3.5). Closed loop transfer function topology based on PI controller.

The open loop transfer function (*OLTF*) with the controllers is written in the following way:

$$OLTF = \frac{\dfrac{p + \dfrac{K_i}{K_p}}{\dfrac{p}{K_p}} * \dfrac{L_m * V_s}{L_s * \left(L_r - \dfrac{L_m^2}{L_s}\right)}}{p + \left(L_r - \dfrac{L_m^2}{L_s}\right)}.$$

(3.59)

By the compensation method: $\dfrac{K_i}{K_p} = L_s * \left(L_r - \dfrac{L_m^2}{L_s} \right)$, we get the next *OLTF*:

$$OLTF = \dfrac{\dfrac{K_p * L_m * V_s}{L_s * \left(L_r - \dfrac{L_m^2}{L_s} \right)}}{p}. \tag{3.60}$$

And the closed loop transfer function (*CLTF*):

$$CLTF = \dfrac{1}{1 + \tau_{cl} * P} \text{ with } \tau_{cl} = \dfrac{\dfrac{1}{K_p} * L_s * \left(L_r - \dfrac{L_m^2}{L_s} \right)}{L_m * V_s}. \tag{3.61}$$

With τ_{cl} is the response time of the system that is fixed at *10 (ms)*. Corresponding to a sufficiently fast value for the use made on the wind turbine and where the variations of the wind are not very fast and the mechanical constants time are important. From **(3.61)**, we can determine the gains[2]K_p *and* K_i according to the parameters of the machine and the response time:

$$K_p = \dfrac{\dfrac{1}{\tau_{cl}} * L_s * \left(L_r - \dfrac{L_m^2}{L_s} \right)}{L_m * V_s} \wedge K_i = \dfrac{\dfrac{1}{\tau_{cl}} * L_s * R_r}{L_m * V_s}. \tag{3.62}$$

4. WIND TURBINE MATHEMATICAL MODEL

The wind turbine[3] input power usually is [12 - 14]:

$$P_v = \dfrac{1}{2} * \rho * S_w * v^3 \tag{3.63}$$

Where ρ is air density; S_w is wind turbine blades swept area in the wind; v is wind speed.

The output mechanical power of wind turbine is:

$$P_m = C_p * P_v = \frac{1}{2} * C_p * \rho * S_w * v^3 \tag{3.64}$$

Where power coefficient[4] (*Cp*) represents the wind turbine power conversion efficiency (*as shown in* (Fig. **3.6**). *Cp* is a non-linear function of *Tip Speed Ratio (TSR or λ)* and the blade angle β [°]. λ is defined as the ratio of the tip speed of the turbine blades to wind speed. λ is given by:

$$\lambda = \frac{R * \Omega_t}{v} \tag{3.65}$$

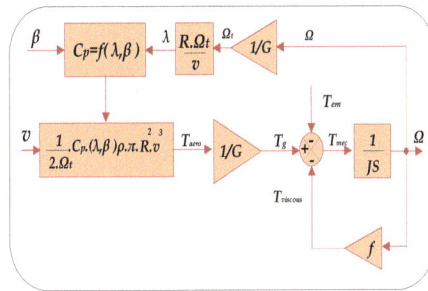

Fig. (3.6). Schematic block of wind turbine.

Where *R* is blade radius, Ω_t is angular speed of the turbine. The expression for C_p (λ, β) can be described as [14]:

$$C_p(\lambda, \beta) = \left(0.5 - 0.0167 * (\beta - 2)\right) * \sin\left[\frac{\pi * (\lambda + 0.1)}{18.5 - 0.3 * (\beta - 2)}\right] - 0.00184 * (\lambda \tag{3.66}$$

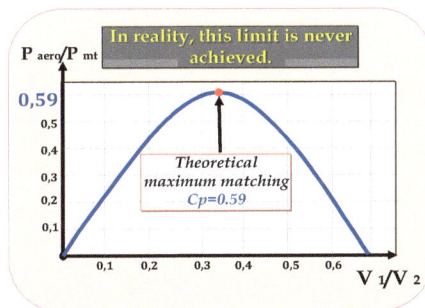

Fig. (3.7). Theorical maximum power coefficient (Cp= (16/27) ≈ 0.59).

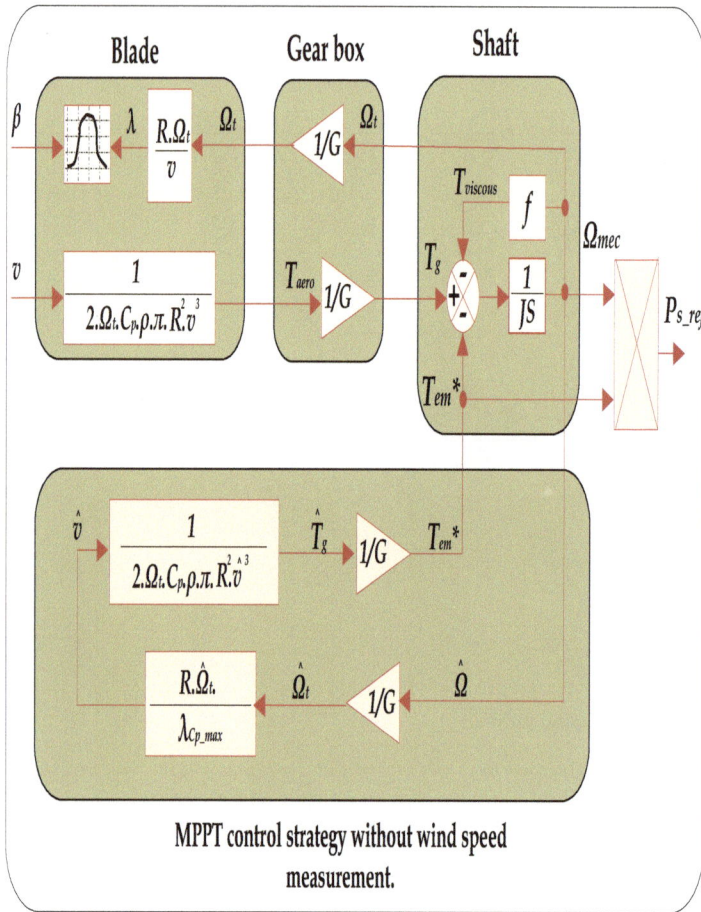

Fig. (3.8). MPPT control strategy without wind speed measurement.

4.1. Maximum Power Point Tracking (MPPT) Strategy

The MPPT strategy without wind speed measurement is illustrates in Fig. (**3.8**); "So the P_{s_ref} (reference stator power) is calculated by the product of the MPPT's output (the mechanical speed Ω_{mec}) by the electromagnetic torque T_{em_ref}"; so $P_{s\_ref}= \Omega_{mec} * T_{em\_ref}$). Using *Matlab/Simulink*® to show the behavior of the Power coefficient (C_p) under different pitch angles (B°) as shown in Fig. (**3.9**), it is clear the zero pitch angle (B°=0°) offered the maximum C_p which correspond to Optimal Tip Speed Ratio (TSR). Fig. (**3.10**) illustrates the three dimensions (3D) of C_p *versus* TSR and pitch angles (B°) respectively. The main aim of the MPPT strategy; is to adapt the speed of the turbine to the wind speed, in order to maximize the converted power, this will improve its energy efficiency and its integration with the electrical grid.

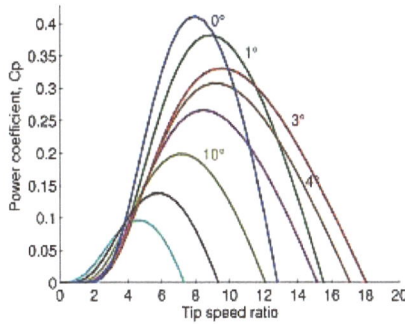

Fig. (3.9). Cp under different pitch angles (B°).

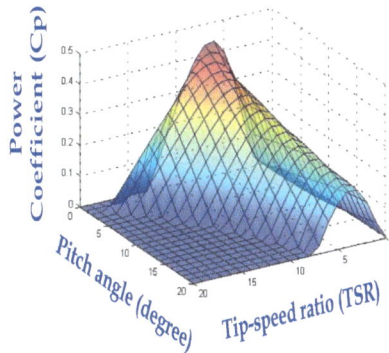

Fig. (3.10). 3D Power coefficient *versus* Tip speed ratio (TSR) and Pitch angle degree (B°).

In this eBook, two wind speed profiles are proposed (step wind speed and random wind speed), as shown in Fig. (**3.11**) (*top side of* Fig. (**3.11**) *present the simulation results of the step wind speed and the bottom side of* Fig. (**3.11**) *illustrates the simulation results of random wind speed*). From the simulation results of Fig. (**3.11**), it can be seen that the maximum value of C_p (*C_{p_max}=0.4785*) is achieved for *B°=0°* and for λ_{Opt}=*8.098*. This point corresponds at the maximum power point tracking (MPPT) [14] as shown in Fig. (**3.11**) (The top-left side of both proposed cases).

After the simulation of the wind turbine model using proposed wind profiles, we test the robustness of the *MPPT* algorithm; we have as results the C_p *versus* time for both proposed wind speed profiles; this latter achieved the maximum value mentioned in Fig. (**3.11**) (*C_{p_max}= 0.4785*) C_{Pmax} = 0.4785 despite the sudden variation of the wind speed. T_{aero}, $T_{gearbox}$ and T_{em} are depicts respectively in Fig. (**3.11**) (*at the bottom*).

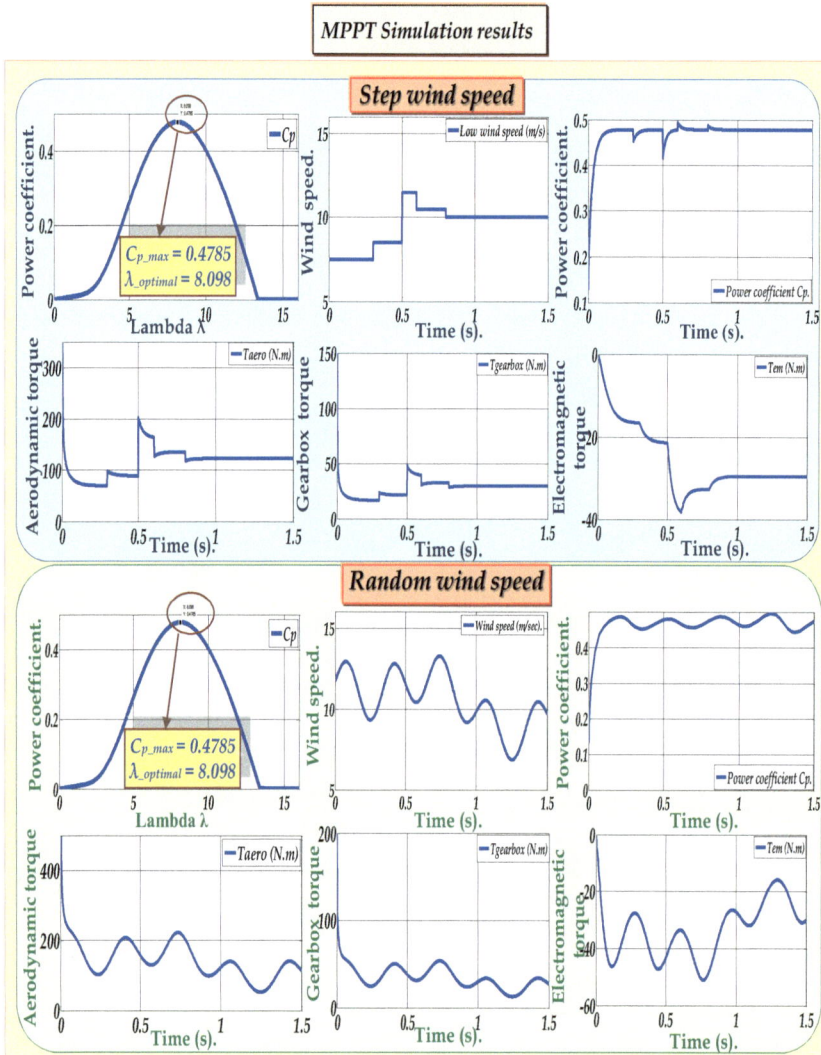

Fig. (3.11). Simulation results of MPPT strategy: (Cp *versus* λ "or TSR", Wind speed *versus* time, Cp *versus* time) Taero, Tgearbox & Tem using two wind speed profiles.

for both proposed wind speed. *Knowing that: T_{em} takes negative value (Generator mode)*. Fig. (**3.12**) illustrates the full circuit topology of a *DFIG* system with a back-to-back *PWM* converter, which is composed of a *GSC* (*Grid Side Converter*), a *RSC* (*Rotor Side Converter*) and a *DC-link* capacitor. Though a few schemes of control, the *DC-link* voltage of the back-to-back *PWM* converter have been studied [26, 27]. In the back-to-back *PWM* converter of *DFIG*, the bidirectional power is transferred between the *GSC* and the *RSC*.

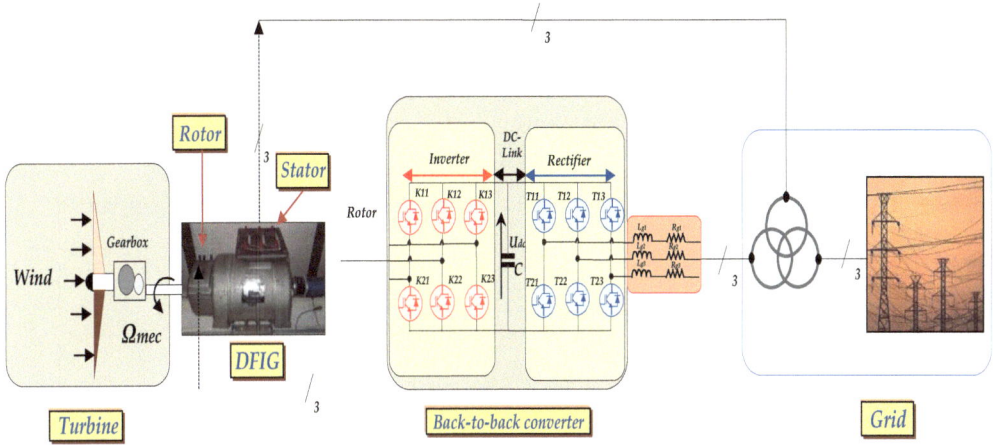

Fig. (3.12). Main circuit topology of a back-to-back PWM converter for DFIG.

5. GRID SIDE CONVERTER (GSC) AND DC-LINK VOLTAGE CONTROL [27 - 29]

Fig. (**3.13**) presents the *GSC* configuration. In this case, we use balanced network voltages, so we will have the following relationships:

$$\begin{cases} V_{1N}=\dfrac{1}{3}*\left(+2*V_1-V_2-V_3\right) \\[2mm] V_{2N}=\dfrac{1}{3}*\left(-V_1+2*V_2-V_3\right) \\[2mm] V_{3N}=\dfrac{1}{3}*\left(-V_1-V_2+2*V_3\right) \end{cases} \tag{3.67}$$

According to the closing or the opening of the switches[6]T_{ij}, the voltages of branch (*leg*) V_i can be equal to V_{dc} or **0**. Other variables such as S_{11}, S_{12} and S_{13} are introduced which take *1* if the switch T_{ij} is closed or *0* if it is blocked. Equation (**3.67**) can be rewritten as:

$$\begin{pmatrix} V_{1N} \\ V_{2N} \\ V_{3N} \end{pmatrix} = \frac{V_{dc}}{3}*\begin{pmatrix} +2 & -1 & -1 \\ -1 & +2 & -1 \\ -1 & -1 & +2 \end{pmatrix}*\begin{pmatrix} S_{11} \\ S_{12} \\ S_{13} \end{pmatrix} \tag{3.68}$$

The rectified current can be written as:

$$I_{rec} = S_{11} * I_{ga} + S_{12} * I_{gb} + S_{13} * I_{gc} \tag{3.69}$$

Where S_{1i} presents a logical signal deduced from the application of the control technique of *PWM*. In this section, the switching signals are determined by the comparison (*using hysteresis controllers*) between the measured grid currents I_{g_abc} and the reference grid currents $I^*_{g\_abc}$.

The terminal voltage of the capacitor is calculated by:

$$\frac{C * dV_{dc}}{dt} = I_c = I_{rec} - I_f = \left(S_{11} * I_{ga} + S_{12} * I_{gb} + S_{13} * I_{gc} \right) - I_f \tag{3.70}$$

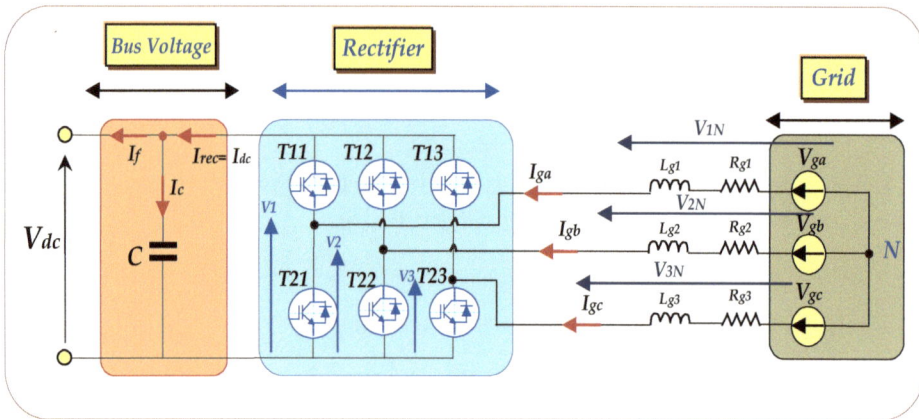

Fig. (3.13). GSC configuration.

Fig. (**3.14**) displays the control block diagram of a vector control strategy for the *GSC*. This grid *PWM converter* is operated to keep the *DC-link* voltage at a constant value. The *GSC* is usually controlled with a vector control strategy with the grid voltage orientation. This voltage frame corresponds to the *d-q axes*, which makes it possible to decouple the expressions from the active and the reactive power exchanged between the *grid* and the *rotor side*. A *DC* capacitor is used in order to remove ripple and keep the *DC-link* voltage relatively smooth. Therefore, a hysteresis controller is used in which the error between the desired and actual currents is passed through a controller [28].

The control of active power and consequent control of the *DC-link* voltage are realized by the intermediary of reference direct grid current I^*_d and the reactive power by the intermediary of reference transversal grid current I^*_{gq}. In order to guarantee the unity power factor ($PF \approx 1$) at the *grid-side*; the reference transversal grid current I^*_{gq} is maintained to zero value ($I^*_{gq} = 0\ (A)$).

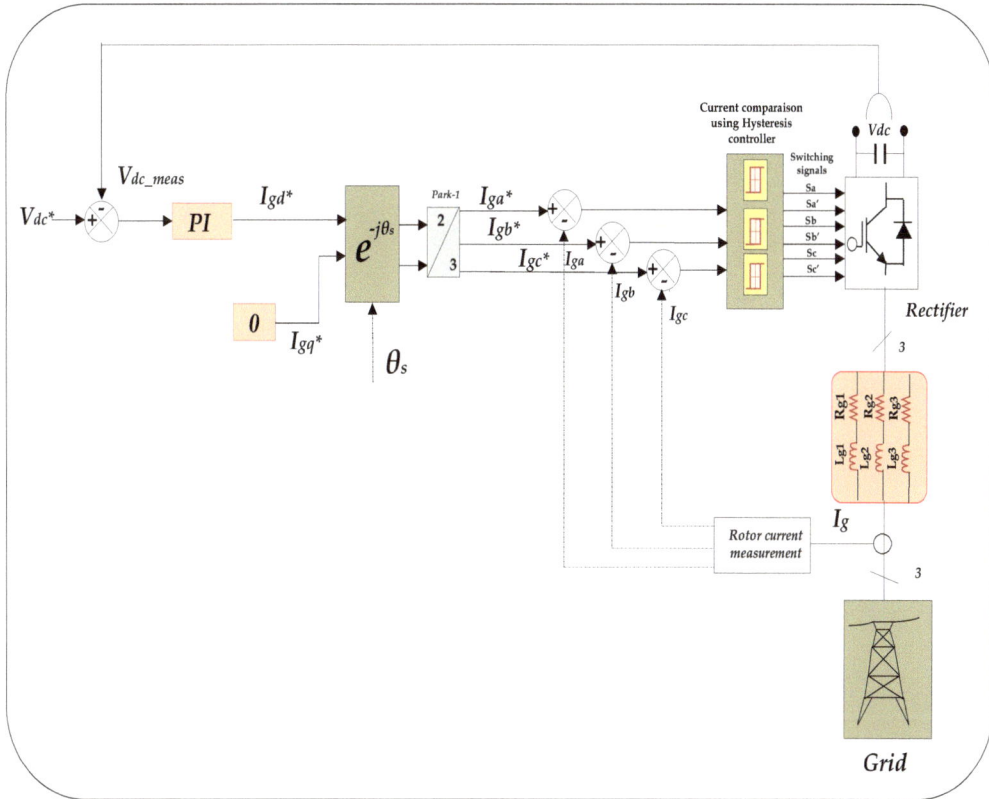

Fig. (3.14). Grid side converter topology (DC-link voltage control).

6. ROTOR SIDE CONVERTER (RSC)

The control of the rotor side converter (*RSC*) enables us to control the stator active and reactive powers independently. From *equations* (**3.49**) *and* (**3.50**), it's clear that the active and reactive powers are based on the *q* and *d axes* rotor currents respectively. Therefore, the powers are checked by controlling the rotor currents. These currents are controlled by *PI* controllers. Then, we must add terms of compensation and decoupling (Fig. **3.15** *illustrates the global RSC configuration*). The voltages obtained are transformed to '*abc*' frame using '*dq*' to '*abc*' transformation which the angle is the difference between the stator angle obtained using *PLL (Phase locked Loop)* and the rotor angle (*as shown in* Fig. (**3.16**)) .

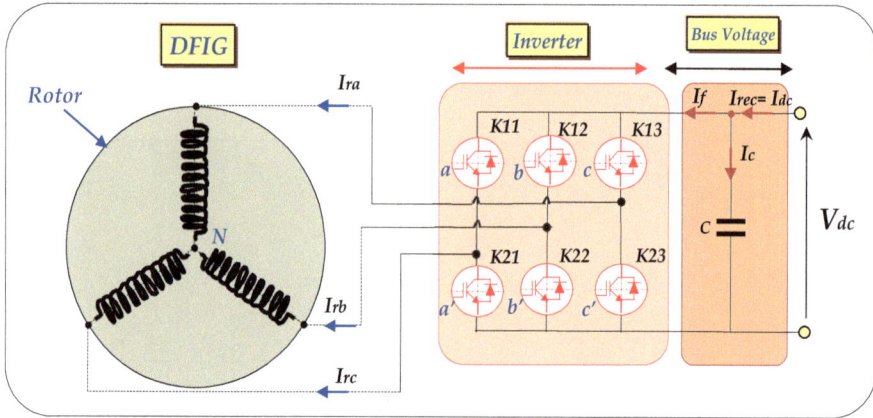

Fig. (3.15). RSC configuration (three-phase voltage source PWM Inverter+ DFIG).

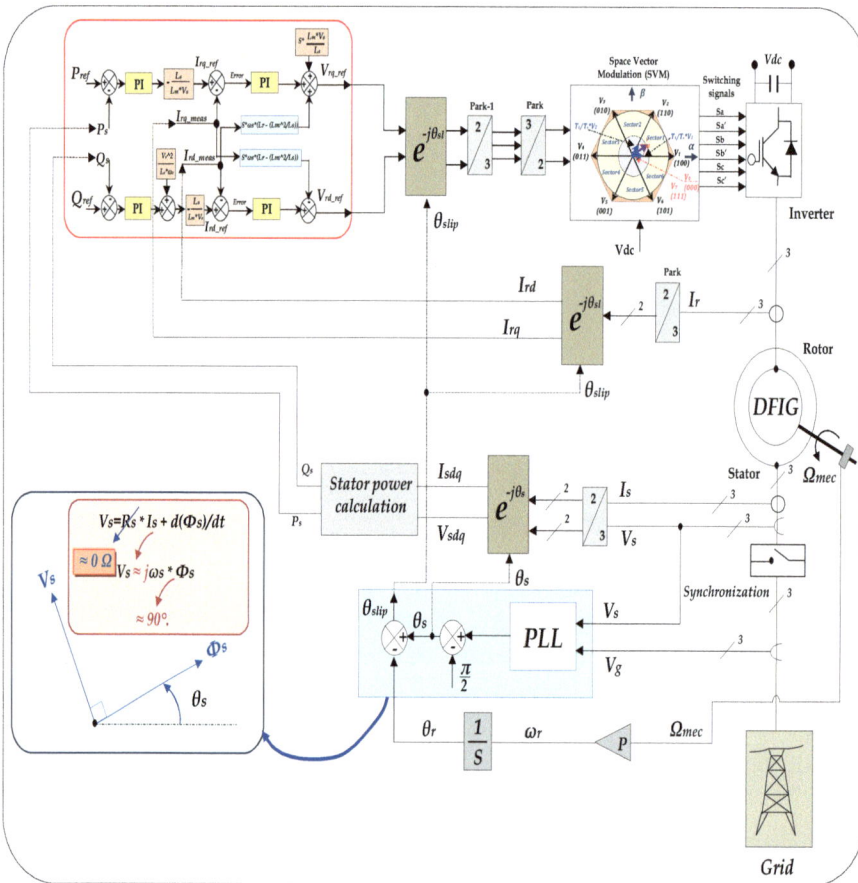

Fig. (3.16). Rotor side converter topology (of Indirect Power Control DFIG).

Finally, the '*abc*' results voltages are transformed to 'Alpha-Beta' and by the means of the DC bus voltage and sectors; it can be converted to SVM (Space vector modulation) signals in order to control the gates of IGBTs used in the rotor side converter (RSC) as set forth in Fig. (**3.15**) [26 - 30].

6.1. Space Vector Modulation (SVM) [31, 32]

The circuit model of a typical three-phase voltage source PWM (Pulse width modulation) inverter is shown in Fig. (**3.15**). K_{11} to K_{23} are the six power switches[6] that shape the output, which are controlled by the switching variables a, a', b, b', c and c'. When an upper transistor is switched on, *i.e.*, when a, b or c is **1**, the corresponding lower transistor is switched off, *i.e.*, the corresponding a', b' or c' is **0**. Therefore, the on and off states of the upper transistors K_{11}, K_{12} and K_{13} can be used to determine the output voltage.

The relationship between the switching variable vector [a, b, c]^t and the line-to-line voltage vector [V_{rab} V_{rbc} V_{rca}]^t is given by equation (**3.71**) in the following:

$$\begin{vmatrix} V_{rab} \\ V_{rbc} \\ V_{rca} \end{vmatrix} = V_{dc} * \begin{vmatrix} +1 & -1 & 0 \\ 0 & +1 & -1 \\ -1 & 0 & +1 \end{vmatrix} * \begin{vmatrix} a \\ b \\ c \end{vmatrix} \tag{3.71}$$

Also, the relationship between the switching variable vector [a, b, c]^t and the phase voltage vector [V_a V_b V_c]^t can be expressed below (equation (**3.72**)).

$$\begin{vmatrix} V_{ran} \\ V_{rbn} \\ V_{rcn} \end{vmatrix} = \frac{V_{dc}}{3} * \begin{vmatrix} +2 & -1 & -1 \\ -1 & +2 & -1 \\ -1 & -1 & +2 \end{vmatrix} * \begin{vmatrix} a \\ b \\ c \end{vmatrix} \tag{3.72}$$

As illustrated in Fig. (**3.17**), there are eight possible combinations of on and off patterns for the three upper power switches. The on and off states of the lower power devices are opposite to the upper one and so are easily determined once the states of the upper power transistors are determined. According to equations (**3.71**) and (**3.72**), the eight switching vectors, output line to neutral voltage (*phase voltage*), and output line-to-line voltages in terms of *DC-link* V_{dc}, are given in Table **3.1** and Fig. (**3.17**) shows the eight inverter voltage vectors (V_0 to V_7).

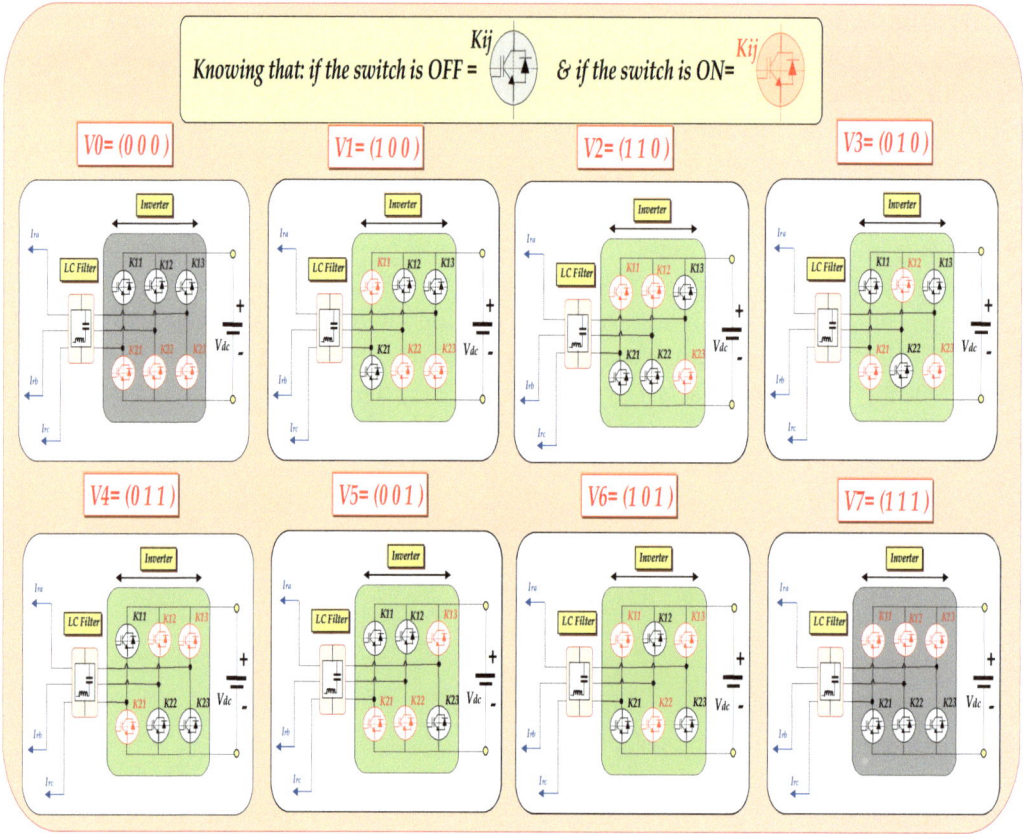

Fig. (3.17). The eight inverter voltage vectors (V_0 to V_7).

Space Vector *PWM* (*SVPWM*) refers to a special switching sequence of the upper three power transistors of a three-phase power inverter. It has been shown to generate less harmonic distortion in the output voltages and or currents applied to the phases of an *AC* motor and to provide more efficient use of supply voltage compared with sinusoidal modulation technique.

Table 3.1. Switching vectors, phase voltages and output line to line voltages.

Voltage Vectors	Switching Vectors			Line to Neutral Voltage			Line to Line Voltage		
	a	b	c	V_{an}	V_{bn}	V_{cn}	V_{ab}	V_{bc}	V_{ca}
V_0	0	0	0	0	0	0	0	0	0
V_1	1	0	0	2/3	-1/3	-1/3	+1	0	-1
V_2	1	1	0	1/3	1/3	-2/3	0	+1	-1
V_3	0	1	0	-1/3	2/3	-1/3	-1	+1	0

(Table 3.1) cont.....

Voltage Vectors	Switching Vectors			Line to Neutral Voltage			Line to Line Voltage		
	a	b	c	V_{an}	V_{bn}	V_{cn}	V_{ab}	V_{bc}	V_{ca}
V_4	0	1	1	-2/3	1/3	1/3	-1	0	+1
V_5	0	0	1	-1/3	-1/3	2/3	0	-1	+1
V_6	1	0	1	1/3	-2/3	1/3	+1	-1	0
V_7	1	1	1	0	0	0	0	0	0

NB: The respective voltage should be multiplied by V_{dc}.

As described in Fig. (**3.18**), this transformation is equivalent to an orthogonal projection of *[a, b, c]*^{t} onto the two-dimensional perpendicular to the vector [1, 1, 1]^t (*the equivalent d-q plane*) in a three-dimensional coordinate system. As a result, six non-zero vectors and two zero vectors are possible. Six non-zero vectors (*V$_1$ - V$_6$*) shape the axes of a hexagonal as depicted in Fig. (**3.19**), and feed electric power to the load. The angle between any adjacent two non-zero vectors is *sixty* degrees (*60°*). Mean-while, two zero vectors (*V$_0$ and V$_7$*) are at the origin and apply zero voltage to the load. The eight vectors are called the basic space

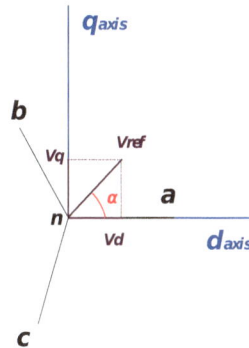

Fig. (3.18). Voltage space vector and its components in d & q axes.

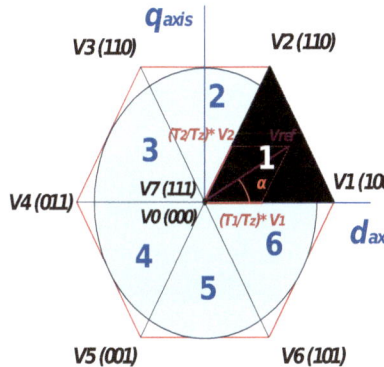

Fig. (3.19). The eight (08) basic switching vectors and sectors.

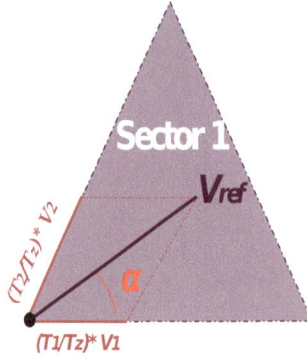

Fig. (3.20). Reference vector as a combination of adjacent vectors at sector 1.

Table 3.2. Switching time calculation at each sector.

Sector	Upper switches (K_{11}, K_{12}, K_{13})	Lower switches (K_{21}, K_{22}, K_{23})
1	$K_{11=}\ T_1 + T_2 + T_0/2$ $K_{12=}\ T_2 + T_0/2$ $K_{13=}\ T_0/2$	$K_{21=}\ T_0/2$ $K_{22=}\ T_1 + T_0/2$ $K_{23=}\ T_1 + T_2 + T_0/2$
2	$K_{11=}\ T_1 + T_0/2$ $K_{12=}\ T_1 + T_2 + T_0/2$ $K_{13=}\ T_0/2$	$K_{21=}\ T_2 + T_0/2$ $K_{22=}\ T_0/2$ $K_{23=}\ T_1 + T_2 + T_0/2$
3	$K_{11=}\ T_0/2$ $K_{12=}\ T_1 + T_2 + T_0/2$ $K_{13=}\ T_2 + T_0/2$	$K_{21=}\ T_1 + T_2 + T_0/2$ $K_{22=}\ T_0/2$ $K_{23=}\ T_1 + T_0/2$
4	$K_{11=}\ T_0/2$ $K_{12=}\ T_1 + T_0/2$ $K_{13=}\ T_1 + T_2 + T_0/2$	$K_{21=}\ T_1 + T_2 + T_0/2$ $K_{22=}\ T_2 + T_0/2$ $K_{23=}\ T_0/2$
5	$K_{11=}\ T_2 + T_0/2$ $K_{12=}\ T_0/2$ $K_{13=}\ T_1 + T_2 + T_0/2$	$K_{21=}\ T_1 + T_0/2$ $K_{22=}\ T_1 + T_2 + T_0/2$ $K_{23=}\ T_0/2$
6	$K_{11=}\ T_1 + T_2 + T_0/2$ $K_{12=}\ T_0/2$ $K_{13=}\ T_1 + T_0/2$	$K_{21=}\ T_0/2$ $K_{22=}\ T_1 + T_2 + T_0/2$ $K_{23=}\ T_2 + T_0/2$

vectors and are denoted by V_0, V_1, V_2, V_3, V_4, V_5, V_6, and V_7. The same transformation can be applied to the desired output voltage to get the desired reference voltage vector V_{ref} in the *d-q* plane. The objective of space vector *PWM* technique is to approximate the reference voltage vector V_{ref} using the eight switching patterns. One simple method of approximation is to generate the average output of the inverter in a small period, *T* to be the same as that of V_{ref} in the same period. Knowing that: the main aim of the SVM strategy is to fix the *IGBT's* switching frequency.

Therefore, space vector PWM can be implemented by the following steps:

<u>Step 1</u>: Determine V_d, V_q, V_{ref}, and angle (α):

From Fig. (**3.18**), the V_d, V_q, V_{ref}, and angle (α) can be determined as follows:

$$V_d = V_{an} - V_{bn} * \cos(60°) - V_{cn} * \cos(60°) = V_{an} - \frac{1}{2} * V_{bn} - \frac{1}{2} * V_{cn}. \tag{3.73}$$

$$V_q = 0 - V_{bn} * \cos(30°) - V_{cn} * \cos(30°) = V_{an} + \frac{\sqrt{3}}{2} * V_{bn} - \frac{\sqrt{3}}{2} * V_{cn}. \tag{3.74}$$

$$\begin{pmatrix} V_d \\ V_q \end{pmatrix} = \frac{2}{3} * \begin{pmatrix} 1 & \frac{-1}{2} & \frac{-1}{2} \\ 0 & \frac{+\sqrt{3}}{2} & \frac{-\sqrt{3}}{2} \end{pmatrix} * \begin{pmatrix} V_{an} \\ V_{bn} \\ V_{cn} \end{pmatrix}. \tag{3.75}$$

$$\left| \overrightarrow{V_{ref}} \right| = \sqrt{V_d^2 + V_q^2}. \tag{3.76}$$

$$a = \tan^{-1}\left(\frac{V_q}{V_d}\right) = \omega.t = 2 * \pi * f. \tag{3.77}$$

Where 'f' is fundamental frequency.

<u>Step 2</u>: Determine time duration T_1, T_2, T_0:

From Fig. (**3.20**), the switching time duration can be calculated as follows:

Switching time duration at Sector 1:

$$\int_0^{T_z} \overrightarrow{V_{ref}} = \int_0^{T_1} \overrightarrow{V_1} dt + \int_{T_1}^{T_1+T_2} \overrightarrow{V_2} dt + \int_{T_1+T_2}^{T_z} \overrightarrow{V_0}. \tag{3.78}$$

$$T_z . \overrightarrow{V_{ref}} = \left(T_1 * \overrightarrow{V_1} + T_2 * \overrightarrow{V_2}\right). \tag{3.79}$$

$$T_z . \left|\overrightarrow{V_{ref}}\right| . \begin{pmatrix} \cos(a) \\ \sin(a) \end{pmatrix} = \frac{T_1 * 2}{3} * V_{dc} * \begin{pmatrix} 1 \\ 0 \end{pmatrix} + \frac{T_2 * 2}{3} * V_{dc} * \begin{pmatrix} \cos\frac{\pi}{3} \\ \sin\frac{\pi}{3} \end{pmatrix} \tag{3.80}$$

Where: $0° \leq a \leq 60°$.

$$T_1 = \frac{T_z * a * \sin\left(\frac{\pi}{3} - a\right)}{\sin\left(\frac{\pi}{3}\right)}.$$ (3.81)

$$T_2 = \frac{T_z * a * \sin(a)}{\sin\left(\frac{\pi}{3}\right)}.$$ (3.82)

$$T_0 = T_z - (T_1 + T_2)$$ (3.83)

Where: $T_z = (1/f_z)$ *and* $a = \left|\overrightarrow{V_{ref}}\right| / \left(\frac{2}{3} * V_{dc}\right).$

Switching time duration at any Sector:

$$T_1 = \frac{\sqrt{3} * T_z * \left|\overrightarrow{V_{ref}}\right|}{V_{dc}} * \left(\sin\left(\frac{\pi}{3} - \alpha + \frac{n-1}{3}\pi\right)\right) = \frac{\sqrt{3} * T_z * \left|\overrightarrow{V_{ref}}\right|}{V_{dc}} *.$$ (3.84)

$$T_2 = \frac{\sqrt{3} * T_z * \left|\overrightarrow{V_{ref}}\right|}{V_{dc}} * \left(\sin\left(\alpha - \frac{n-1}{3}\pi\right)\right) = \frac{\sqrt{3} * T_z * \left|\overrightarrow{V_{ref}}\right|}{V_{dc}} * \left(-\cos(\alpha).\sin\right.$$ (3.85)

$$T_0 = T_z - (T_1 + T_2)$$ (3.86)

Where, n=1 through 6 (that is: Sector 1 to 6) and $0° \leq a \leq 60°$.

<u>Step 3</u>: Determine the switching time of each transistor (K_{11} to K_{23})

Fig. (**3.21**) shows space vector PWM switching patterns at each sector. The switching time at each sector is summarized in Table **3.2**, and it will be built in Simulink model to implement SVM.

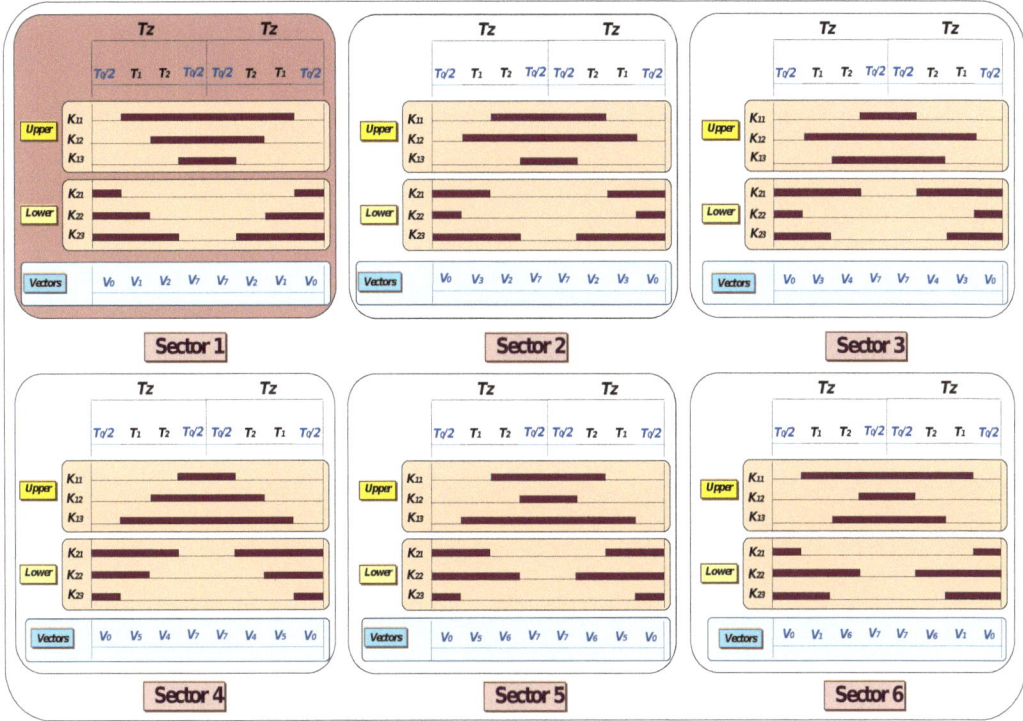

Fig. (3.21). Space vector PWM switching patterns at each sector.

6.2. LC Filter

Fig. (**3.22**) shows *L-C* output filter to obtain current and voltage equations. By applying Kirchhoff's current law to nodes a, b, and c, respectively, the following current equations are derived:

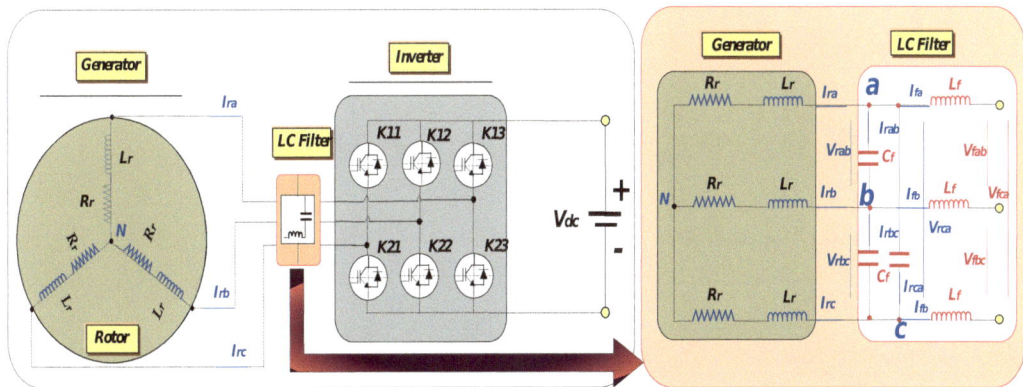

Fig. (3.22). Rotor side converter topology with LC-Filter (of Indirect Power Control DFIG).

Node "1":

$$I_{fa}+I_{ca}=I_{ab}+I_{ra}\Rightarrow I_{fa}+C_f.\frac{dV_{rca}}{dt}=C_f.\frac{dV_{rab}}{dt}+I_{ra}. \quad (3.87)$$

Node "2":

$$I_{fb}+I_{ab}=I_{bc}+I_{rb}\Rightarrow I_{fb}+C_f.\frac{dV_{rab}}{dt}=C_f.\frac{dV_{rbc}}{dt}+I_{rb}. \quad (3.88)$$

Node "3":

$$I_{fc}+I_{bc}=I_{ca}+I_{rc}\Rightarrow I_{fa}+C_f.\frac{dV_{rbc}}{dt}=C_f.\frac{dV_{rca}}{dt}+I_{rc}. \quad (3.89)$$

Where:

$$I_{ab}=C_f.\frac{dV_{rab}}{dt}, \; I_{bc}=C_f.\frac{dV_{rbc}}{dt} \text{ and } I_{ca}=C_f.\frac{dV_{rca}}{dt} \quad (3.90)$$

Also, equations: **(3.87)** to **(3.89)** can be rewritten as the following equations, respectively:

1/subtracting equations: **(3.88)** from **(3.87)**:

$$I_{fa}-I_{fb}+C_f.\left(\frac{dV_{rca}}{dt}-\frac{dV_{rab}}{dt}\right)=C_f.\left(\frac{dV_{rab}}{dt}-\frac{dV_{rbc}}{dt}\right)+I_{ra}-I_{rb}$$
$$\Rightarrow C_f.\left(\frac{dV_{rca}}{dt}+\frac{dV_{rbc}}{dt}-2.\frac{dV_{rab}}{dt}\right)=-I_{fa}+I_{fb}+I_{ra}-I_{rb}. \quad (3.91)$$

2/subtracting equations: **(3.89)** from **(3.88)**:

$$I_{fb}-I_{fc}+C_f.\left(\frac{dV_{rab}}{dt}-\frac{dV_{rbc}}{dt}\right)=C_f.\left(\frac{dV_{rbc}}{dt}-\frac{dV_{rca}}{dt}\right)+I_{rb}-I_{rc}$$
$$\Rightarrow C_f.\left(\frac{dV_{rab}}{dt}+\frac{dV_{rca}}{dt}-2.\frac{dV_{rbc}}{dt}\right)=-I_{fb}+I_{fc}+I_{rb}-I_{rc}. \quad (3.92)$$

3/subtracting equations: **(3.87)** from **(3.89)**:

$$I_{fc} - I_{fa} + C_f \cdot \left(\frac{dV_{rbc}}{dt} - \frac{dV_{rca}}{dt} \right) = C_f \cdot \left(\frac{dV_{rca}}{dt} - \frac{dV_{rab}}{dt} \right) + I_{rc} - I_{ra}$$

$$\Rightarrow C_f \cdot \left(\frac{dV_{rab}}{dt} + \frac{dV_{rbc}}{dt} - 2 \cdot \frac{dV_{rca}}{dt} \right) = -I_{fc} + I_{fa} + I_{rc} - I_{ra}. \tag{3.93}$$

To simplify equations: **(3.91)** to **(3.93)**, we use the following relationship that an algebraic sum of line to line rotor voltages is equal to zero:

$$V_{rab} + V_{rbc} + V_{rca} = 0. \tag{3.94}$$

Based on **(3.94)**, the **(3.91)** to **(3.93)** can be modified to a first-order differential equation, respectively:

$$\left| \begin{array}{l} \dfrac{dV_{rab}}{dt} = \dfrac{1}{3.C_f} * I_{fab} - \dfrac{1}{3.C_f} * I_{rab} \\[3mm] \dfrac{dV_{rbc}}{dt} = \dfrac{1}{3.C_f} * I_{fbc} - \dfrac{1}{3.C_f} * I_{rbc}. \\[3mm] \dfrac{dV_{rca}}{dt} = \dfrac{1}{3.C_f} * I_{fca} - \dfrac{1}{3.C_f} * I_{rca} \end{array} \right. \tag{3.95}$$

Where: $I_{fab} = I_{fa} - I_{fb}$, $I_{fbc} = I_{fb} - I_{fc}$, $I_{fca} = I_{fc} - I_{fa}$ and $I_{rab} = I_{ra} - I_{rb}$, $I_{rbc} = I_{rb} - I_{rc}$, $I_{rca} = I_{rc} - I_{ra}$.

By applying Kirchhoff's voltage law on the side of inverter output, the following voltage equations can be derived:

$$\left| \begin{array}{l} \dfrac{dI_{fab}}{dt} = \dfrac{1}{L_f} * V_{rab} + \dfrac{1}{L_f} * V_{fab} \\[3mm] \dfrac{dI_{fbc}}{dt} = \dfrac{1}{L_f} * V_{rbc} + \dfrac{1}{L_f} * V_{fbc}. \\[3mm] \dfrac{dI_{fca}}{dt} = \dfrac{1}{L_f} * V_{rca} + \dfrac{1}{L_f} * V_{fca} \end{array} \right. \tag{3.96}$$

By applying Kirchhoff's voltage law on the rotor side, the following voltage equations can be derived: Equation **(3.97)** can be rewritten as:

$$
\left\{
\begin{aligned}
V_{rab} &= L_r \frac{d I_{ra}}{dt} + R_r * I_{ra} - L_r \frac{d I_{rb}}{dt} - R_r * I_{rb} \\
V_{rbc} &= L_r \frac{d I_{rb}}{dt} + R_r * I_{rb} - L_r \frac{d I_{rc}}{dt} - R_r * I_{rc} \, . \\
V_{rca} &= L_r \frac{d I_{rc}}{dt} + R_r * I_{rc} - L_r \frac{d I_{ra}}{dt} - R_r * I_{ra}
\end{aligned}
\right.
\tag{3.97}
$$

Equation **(3.97)** can be rewritten as:

$$
\left\{
\begin{aligned}
\frac{d I_{rab}}{dt} &= \frac{-R_r}{L_r} * I_{rab} + \frac{1}{L_r} * V_{rab} \\
\frac{d I_{rbc}}{dt} &= \frac{-R_r}{L_r} * I_{rbc} + \frac{1}{L_r} * V_{rbc} \, . \\
\frac{d I_{rca}}{dt} &= \frac{-R_r}{L_r} * I_{rca} + \frac{1}{L_r} * V_{rca}
\end{aligned}
\right.
\tag{3.98}
$$

Therefore, we can rewrite **(3.95)**, **(3.96)** and **(3.98)** into a matrix form, respectively:

$$
\left\{
\begin{aligned}
\frac{d V_r}{dt} &= \frac{1}{3 * C_f} * I_f - \frac{1}{3 * C_f} * I_r \\
\frac{d I_f}{dt} &= \frac{-1}{L_f} * V_r + \frac{1}{L_f} * V_f \\
\frac{d I_r}{dt} &= \frac{1}{L_r} * V_r - \frac{R_r}{L_r} * I_r
\end{aligned}
\right.
\tag{3.99}
$$

Where: $\mathbf{V}_r = [V_{rab} \ V_{rbc} \ V_{rca}]^T$, $\mathbf{I}_f = [I_{fab} \ I_{fbc} \ I_{fca}]^T = [I_{fa}\text{-}I_{fb} \ I_{fb}\text{-}I_{fc} \ I_{fc}\text{-}I_{fa}]^T$, $\mathbf{V}_f = [V_{fab} \ V_{fbc} \ V_{fca}]^T$, $\mathbf{I}_r = [I_{rab} \ I_{rbc} \ I_{rca}]^T = [I_{ra}\text{-}I_{rb} \ I_{rb}\text{-}I_{rc} \ I_{rc}\text{-}I_{ra}]^T$.

Finally, the given plant model **(3.99)** can be expressed as the following continuous-time state space equation:

$$\dot{x}(t) = A * x(t) + B * u(t) \tag{3.100}$$

Where:

$$x = \begin{bmatrix} V_r \\ I_f \\ I_r \end{bmatrix}_{9*1} \quad A = \begin{bmatrix} 0_{3*3} & \dfrac{1}{3.C_f}.I_{3*3} & \dfrac{-1}{3.C_f}.I_{3*3} \\ \dfrac{-1}{L_f}.I_{3*3} & 0_{3*3} & 0_{3*3} \\ \dfrac{1}{L_{rotor}}.I_{3*3} & 0_{3*3} & \dfrac{-R_{rotor}}{L_{rotor}}.I_{3*} \end{bmatrix} \quad B = \begin{bmatrix} 0_{3*3} \\ \dfrac{1}{L_f}.I_{3*3} \\ 0_{3*3} \end{bmatrix}_{9*3} ,$$

$$u = \begin{bmatrix} V_f \end{bmatrix}_{3*1}$$

7. OPERATING PRINCIPLE OF DFIG

In spite of the disadvantages associated with the slip-rings, the wound-rotor induction machine has long been a wind electric generator choice. By using a suitable integrated approach in the design of a *WECS*, use of a slip-ring induction generator has been found to be economically competitive. Control of grid-connected and isolated variable-speed wind turbines with a doubly fed induction generator has been implemented and reported [33, 34].

A wound-rotor induction machine can be operated as a doubly-fed induction machine (*DFIM*) when a power converter is present in its rotor circuit. This converter directs the power flow into and out of the rotor windings. Because the *DFIM* can operate as either a motor or a generator at sub-synchronous and super-synchronous speeds[8], there exist four operational modes in which the *DFIM* operates. All the four modes are explained in Fig. (**3.23**). When the machine runs above synchronous speed, this operation is termed super-synchronous operation. Similarly, operation below synchronous speed is called sub-synchronous operation. In both sub- and super-synchronous operation, the machine can be operated either as a motor or a generator. In the motoring mode of operation, the torque produced by the machine is positive. On the other hand, during generating operation, the machine needs mechanical torque as input; thus, the torque is negative during generating operation. The principle of a DFIM control in these modes can be understood more clearly by the power-flow diagram given in Fig. (**3.25**). In this figure, P_s is the stator power, P_r is the rotor power, and P_m is the mechanical power. When the *DFIM* is operating as a motor in the sub-synchronous speed range (Fig. (**3.25**)- [1]), power is taken out of the rotor. This operational mode is commonly known as slip-power recovery. If the speed increases so that the machine is operating at super-synchronous speeds (Fig. **3.25** [2]), the rotor power then changes direction [33, 34].

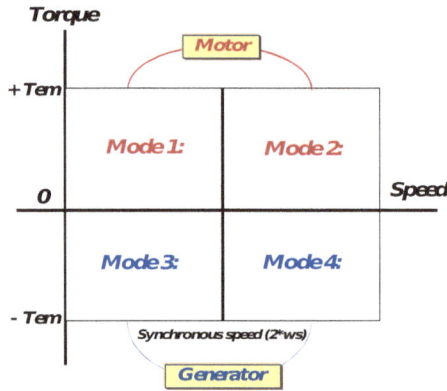

Fig. (3.23). Operating modes of a DFIM: Mode 1 (Sub-synchronous motoring mode), Mode 2 (Super-synchronous motoring mode), Mode 3 (Subsynchronous generating mode), and Mode 4 (Super-synchronous generating mode).

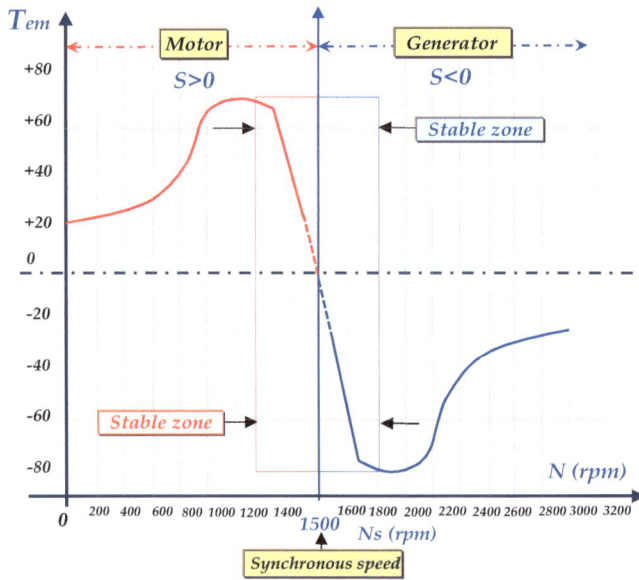

Fig. (3.24). DFIM's torque/speed characteristic in 04 quadrants.

When the *DFIM* is operating as a generator in the sub-synchronous speed range (Fig. **3.25**)- [3]), power is delivered to the rotor. If the speed increases so that the machine is operating at super-synchronous speeds (Fig. **3.25**)- [4]) the rotor power again changes direction. Because the machine will be predominantly working as a generator for a wind-energy application, operation in *Mode 3* and *Mode 4* is more important than *Mode 1* and *Mode 2*. However, for an ideal *WECS* system, all four operating modes are desirable. Motoring modes are useful when the generator needs to speed up quickly in order to achieve the best operating speed and efficiency [33]. However, because of the high inertia of the wind generator power train, the acceleration of the machine may be achieved by the wind torque itself. Hence, motoring operations may be sacrificed if the cost of the system can be reduced substantially. Fig. (**3.24**) displays the *Torque/Speed* characteristic of *DFIM* in four (04) quadrants under both modes: *Motoring and Generating*. It should be noted that for Sub- and Super-synchronous generating modes, the power flows through the rotor are of opposite directions. Hence, the power converter connected with such system should have bidirectional power flow capability.

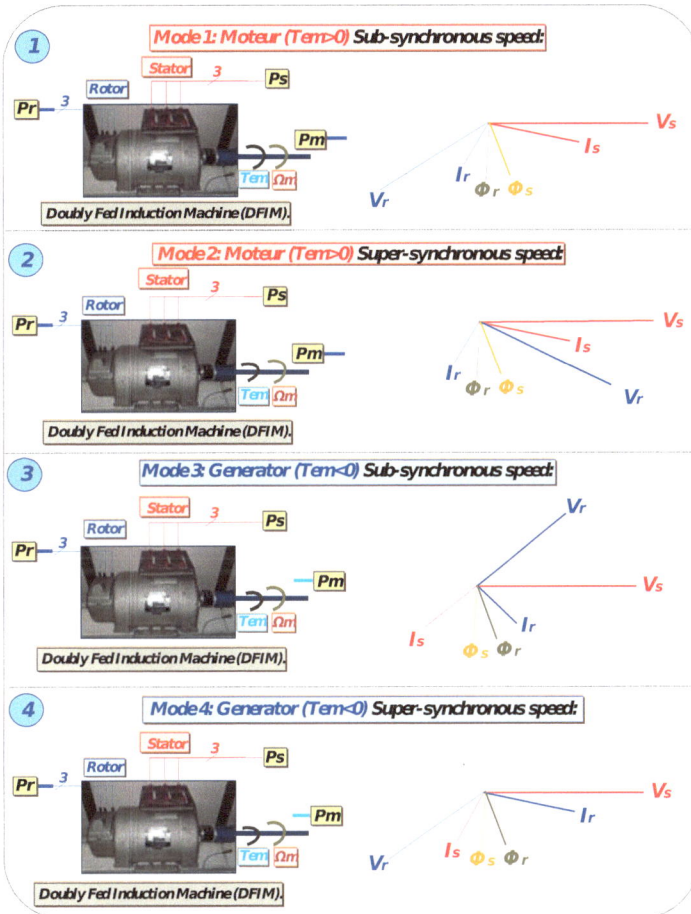

Fig. (3.25). Power-flow diagram of a DFIM for (1): Sub-synchronous motoring mode, (2): Super-synchronous motoring mode, (3): Sub-synchronous generating mode, and (4): Super-synchronous generating mode.

8. EXPERIMENTAL RESULTS OF CLASSICAL POWER CONTROL UNDER SUB-SYNCHRONOUS & SUPER-SYNCHRONOUS OPERATIONS

In this section, a demonstration of Sub- and Super-synchronous operations is presented in (Figs. **3.26** and **3.27**). Fig. (**3.27**)[8] explains the relationship between: electromagnetic torque, synchronous speed and mechanical speed values in four modes (especially in generator modes/ under negative). Figs. (**3.26a,b**) illustrate the experimental results of the rotor sinusoidal current variation in transient and steady states (I_{ra_meas} (A)) phase 'A' and the slip angle variation (θ_{slip} (rad)) in transient and steady states under in the period of 40 (sec)=(4 (sec)*10). It can be seen that the variation of I_{ra_meas} (A) has no effect because the value of slip angle theoretically based on rotor angle and grid angle (knowing that in this case the rotor angle equals to zero; means that the slip angle equal directly to grid angle because the rotor speed Ωr (rad/sec) =N_r (rpm)= 0 (rpm). Figs. (**3.26c,d**) present the behavior of the rotor measured current waveform in transient and steady states in the period of 100 (sec) = (10 (sec)*10) (I_{ra_meas} (A); knowing that the peak to peak current magnitude equals to 10 (A) under rotor speed variation (The variation of Nr (rpm): 0, 1500, 1700, 1000, 1500, 1700, 1000, 500 and 0 respectively); knowing that 1500 (rpm) presents the synchronous speed.

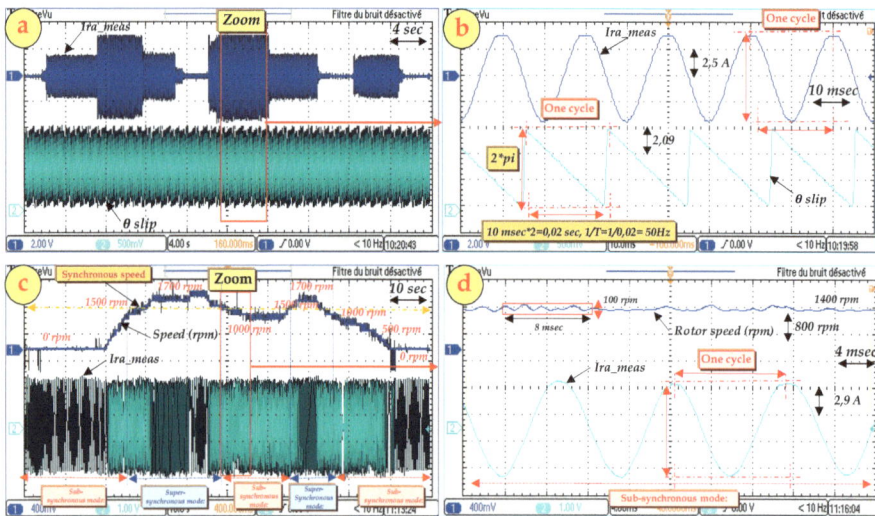

Fig. (3.26). a and **b** Measured rotor current variation and slip angle, **c** and **d**: Rotor speed variation and measured rotor current.

9. PROPOSED IDPC BASED ON PID CONTROLLERS

Proportional-Integral-Derivative (PID controller[9]) is widely used in industrial control system. PID controller has all the necessary dynamics: fast reaction on change of the controller input (D controller), increase in control signal to lead error towards zero (I controller) and suitable action inside control error area to eliminate oscillations (P controller). Derivative mode improves stability of the system and enables increase in gain K_p, which increases speed of the controller response. The output of PID controller consists of three terms the error signal, the error integral and the error derivative. The error signal is computed by equation **(3.101).** Fig. (**3.28**) shows the block diagram of PID controller. The transfer function of PID controller is expressed as [35]:

$$\frac{Y}{Y_{ref}} = K_p + \frac{K_i}{p} + K_d \cdot p \qquad (3.101)$$

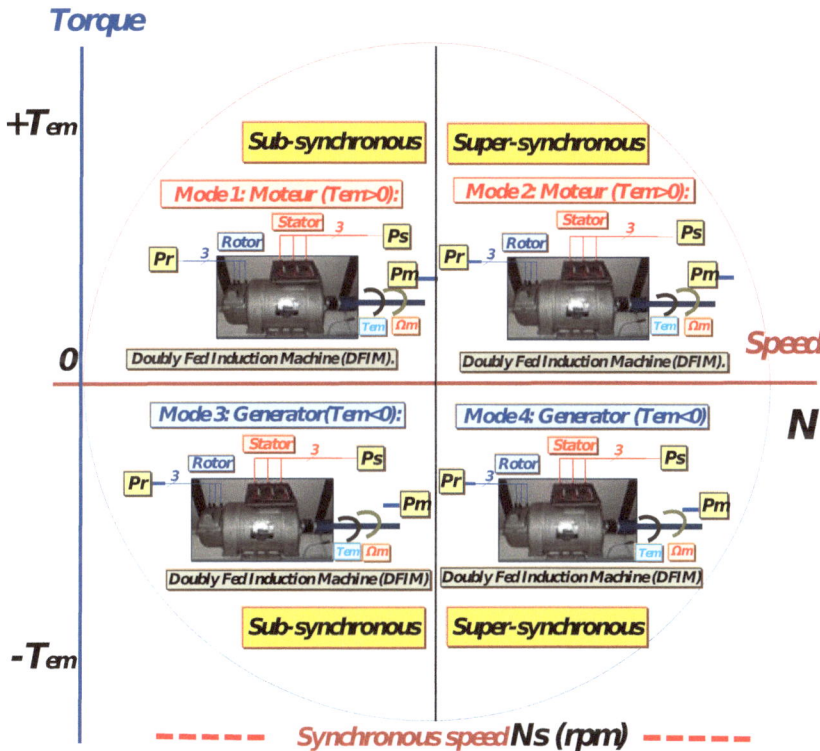

Fig. (3.27). DFIG's operating modes (Motor/Generator) using Doubly Fed Induction Machine (DFIM).

Where K_p is the proportional gain, K_d is the derivative gain, K_i is the integral gain, and Y(s) is the output control signal which represent stator active and reactive power (P_s and Q_s) respectively. Fig. (**3.29**) illustrates the proposed IDPC scheme

(The inputs are = P_s, Q_s, and the output are= V_{rd}, V_{rq}).

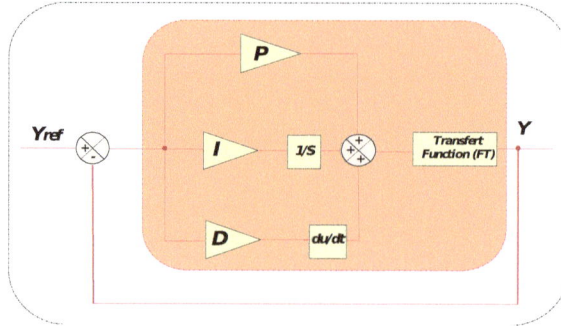

Fig. (3.28). Proportional-Integral-Derivative controller structure.

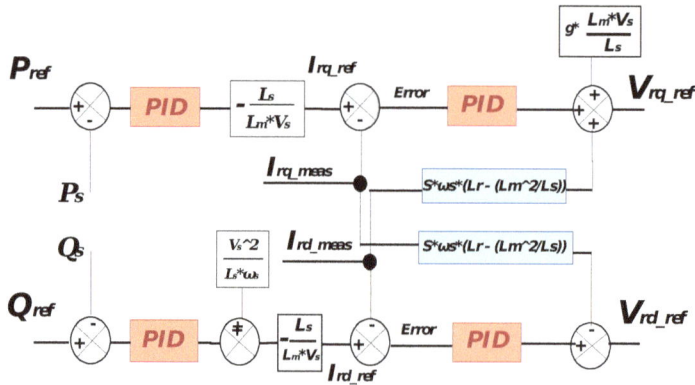

Fig. (3.29). Proposed IDPC (based on PID) of DFIG.

9.1. Advantages

- The PID controller feasibility is easy to be implemented (simple & easy to use it).
- The PID gains can be designed based upon the system parameters if they can be achieved or estimated precisely.
- Less oscillation and low overshoot & faster dynamic response.
- Robust: insensitive to changes to plant parameter & disturbance.

9.2. Drawbacks

- Very sensible to the parameters changing (as the robustness tests).
- Need to the robust mechanism (as Intelligent algorithms: Fuzzy logic & Neuro-Fuzzy Controls), in order to predict the future behavior of wind turbine DFIG.

10. PROPOSED IDPC BASED ON MRAC CONTROLLERS

10.1. Definition

An adaptive controller is a controller that can modify its behavior in response to the changes in dynamics of the processes and the disturbances acting on the process.

10.2. Description

- An adaptive control system can be thought of as having two loops.
- One loop is a normal feedback with the process and the controller.
- The other loop is the parameter adjustment loop.
- The parameter adjustment loop is usually slower than the normal feedback loop.
- An adaptive controller a controller with adjustable parameters and a mechanism for adjusting the parameters.
- The parameters are adjusted to compensate for the changes in dynamics of the plant and the disturbances acting on the plant.
- The controller becomes nonlinear because of the parameter adjustment mechanism.

10.3. Some Mechanisms Causing Variation in Process Dynamics Are

- Nonlinear actuators.
- Flow and speed variations.
- Flight control.
- Variation in disturbance characteristics.

10.4. Advantages

- Parameters can be changed quickly in response to changes in plant dynamics.
- Very easy to apply.

10.5. Drawbacks

- It is an open-loop adaptation scheme, with no real learning or intelligence.
- The design required for its implementation is enormous.

The system studied in this chapter is based on a first-order linear plant approximation given by [20]:

$$\dot{x}(t) = -a * x(t) + b * u(t) \qquad (3.102)$$

Where $x(t)$ is the plant state, $u(t)$ is the control signal and a and b are the plant parameters. The control signal is generated from both the state variable and the reference signal $r(t)$, multiplied by the adaptive control gains K and Kr such that:

$$u(t) = K(t) * x(t) + K_r(t) * r(t) \qquad (3.103)$$

Where $K(t)$ is the feedback adaptive gain and $Kr(t)$ the feed forward adaptive gain. The plant is controlled to follow the output from a reference model.

$$\dot{x}_m(t) = a_m * x_m(t) + b_m * r(t) \qquad (3.104)$$

Where x_m is the state of the reference model and a_m and b_m are the reference model parameters which are specified by the controller designer. The object of the *MRAC* algorithm is for $x_e \to 0 \, as \, t \to \infty$, where $x_e = x_m \text{-} x$ is the error signal. The dynamics of the system may be rewritten in terms of the error such that:

$$\dot{x}_e(t) = a_m * x_e(t) + \left(a - a_m - b * K(t)\right) * x(t) + \left(b_m - b * K_r(t)\right) * r(t) \qquad (3.105)$$

Using Equations **(3.102)**, **(3.103)** and **(3.104)**, it can be seen that for exact matching between the plant and the reference model, the following relations hold.

$$K = K^E = \frac{a - a_m}{b} \qquad (3.106)$$

$$K_r = K_r^E = \frac{b_m}{b} \qquad (3.107)$$

Where $()^E$ denotes the (constant) Erzberger gains [20].

Equations **(3.105)** and **(3.106)** can be used as:

$$\dot{x}_e(t) = - a_m * x_e + b * \left(K^E - K\right) * \left(x_m - x_e\right) + b * \left(K^E - K\right) * r \qquad (3.108)$$

For general model reference adaptive control, the adaptive gains are commonly defined in a proportional plus integral formulation.

$$K\left(e I_{rd_{ref}}, t\right) = \int_0^t a * y_e * I_{rd_{ref}}^T \, dt + b * y_e * I_{rd_{ref}}^T. \qquad (3.109)$$

$$K_r\left(e\,I_{rd},t\right)=\int_0^t a*y_e*I_{rd}^T dt+b*y_e*I_{rd}^T. \tag{3.110}$$

$$K\left(e\,I_{rq_{ref}},t\right)=\int_0^t a*y_e*I_{rq_{ref}}^T dt+b*y_e*I_{rq_{ref}}^T. \tag{3.111}$$

$$K_r\left(e\,I_{rq},t\right)=\int_0^t a*y_e*I_{rq}^T dt+b*y_e*I_{rq}^T. \tag{3.112}$$

Where α and β are adaptive control weightings representing the adaptive effort. y_e is a scalar weighted function of the error state and its derivatives, $y_e = C_e * x_e$, where C_e can be chosen to ensure the stability of the feed forward block.

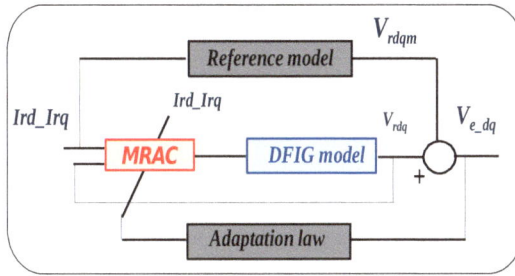

Fig. (3.30). Proposed MRAC structure to control Ird and Irq respectively.

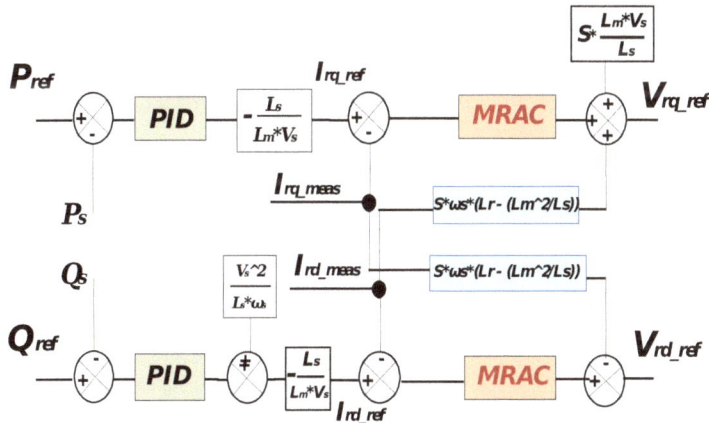

Fig. (3.31). Proposed IDPC (based on MRAC) of DFIG.

In this case, the MRAC controller Fig. (**3.30**) is proposed to control rotor direct and quadrature currents respectively (I_{rd} and I_{rq}) instead the PID controllers, in order to improve the dynamic responses against parameter variation. The simulation results of the proposed algorithms will presented in Figs. (**3.31, 3.33, 3.34 & 3.36**).

Fig. (**3.32**) presents the whole proposed IDPC control which is composed of six (06) important part as follows; **1**: Back-to-back two level converter (based on SVM strategy), **2**: The proposed power profiles, **3**: The MPPT strategy under three (03) proposed mode, **4**: SVM modulation strategy, **5**: proposed IDPC, **6**: PID block diagram and **7**: MRAC block diagram.

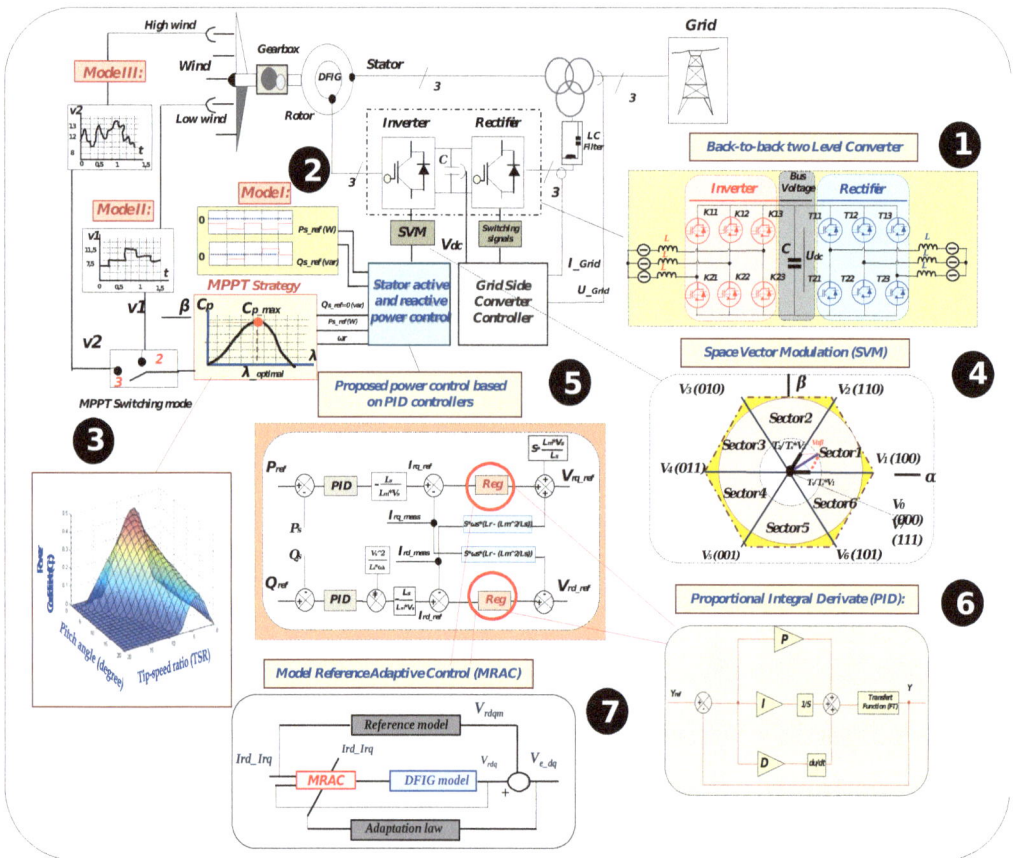

Fig. (3.32). The proposed wind turbine-DFIG algorithm (based on PID & MRAC) using 3 MPPT modes.

11. SIMULATION RESULTS

The proposed system (*DFIG control + wind turbine-refer* to Figs. **A2 & A3** respectively) is validated using *Matlab/Simulink®* software under MPPT strategy (refer to Fig."**A.1**") by keeping stator reactive power equals to zero and to ensure unity power factor (PF=1). Knowing that the simulation parameters are illustrated in Table **A.3** (refer to Appendix). The Figs. (**3.33** to **3.41**) respectively present the simulation results for conventional and proposed IDPC (refer to Fig."**A.3**") using PI, PID & MRAC respectively under two level converters-knowing that the wind-system is based on DFIG (4 kW) and wind turbine (4.5 kW). These figures are divided into three parts;

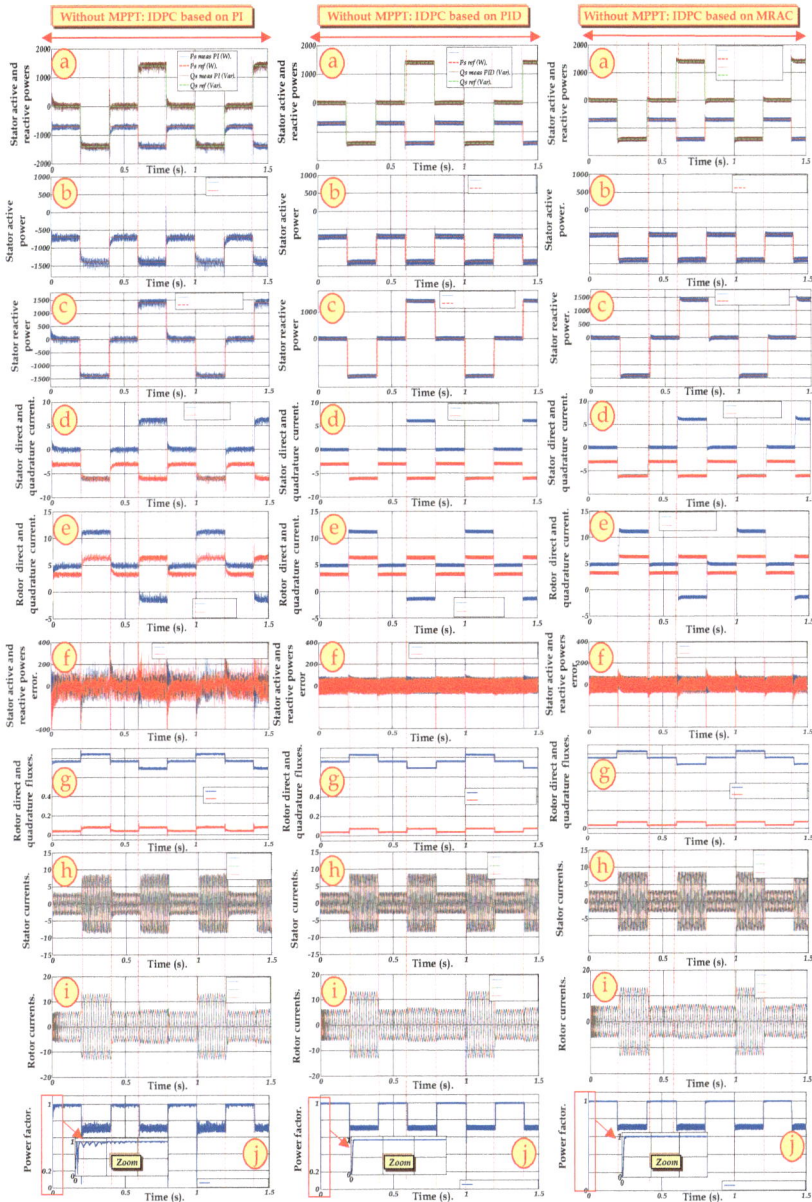

Fig. (3.33). Simulation results of **Mode-1** using PI, PID & MRAC; (**a**): stator active and reactive powers, (**b**): stator active power, (**c**): stator reactive power, (**d**): stator direct and quadrature currents, (**e**): rotor direct and quadrature currents, (**f**): stator active and reactive power error, (**g**): rotor direct and quadrature fluxes, (**h**): stator currents, (**i**): rotor currents, (**j**): power factor.

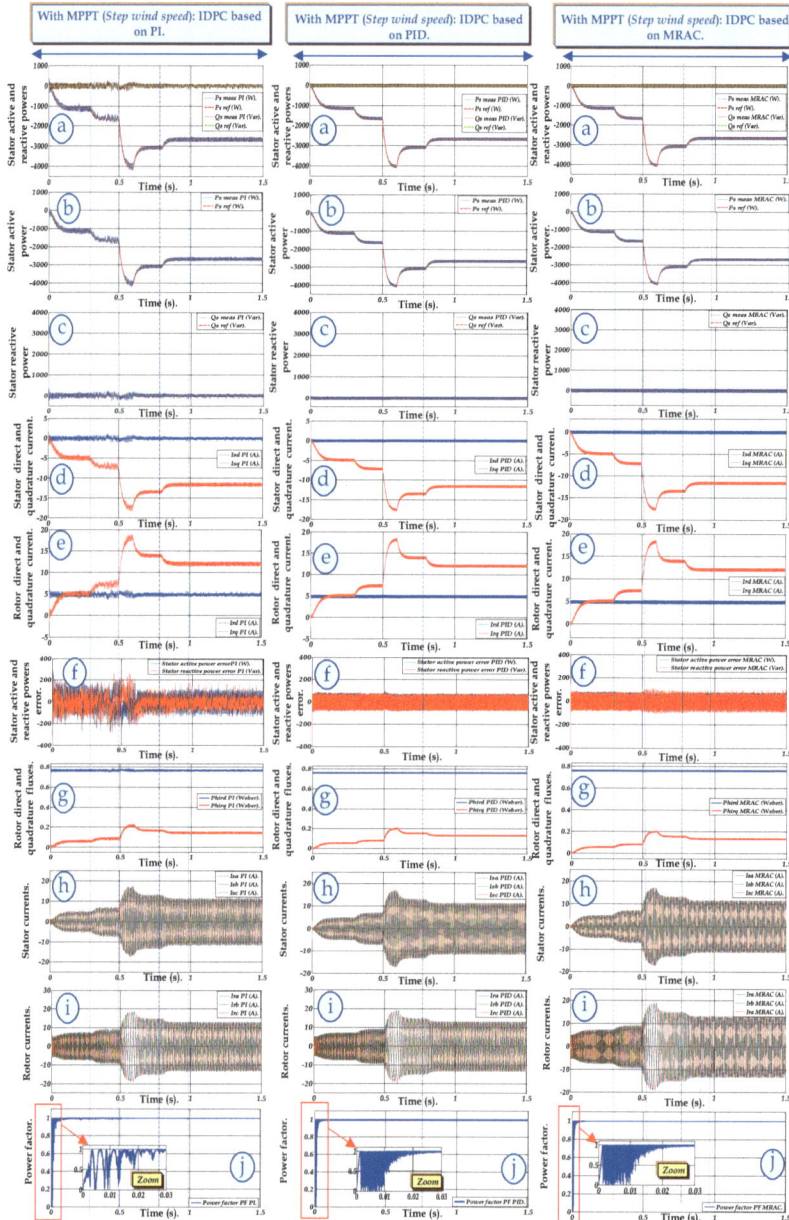

Fig. (3.34). Simulation results of **Mode-2** using PI, PID & MRAC; (**a**): stator active and reactive powers, (**b**): stator active power, (**c**): stator reactive power, (**d**): stator direct and quadrature currents, (**e**): rotor direct and quadrature currents, (**f**): stator active and reactive power error, (**g**): rotor direct and quadrature fluxes, (**h**): stator currents, (**i**): rotor currents, (**j**): power factor.

● The first part depicts the behavior of the wind-system parameters under three (03) modes –*knowing that for each mode we present the differents proposed topologies*-as follows:

1- Mode 1 (Red color/ Fig. (**3.33**): Without MPPT Strategy, in this case we impose the P_s and Q_s reference profiles.

2- Mode 2 (Blue color/ Fig. **3.34**): With MPPT strategy, in this case we propose a low wind speed based on step form (Max wind speed = 11.5 m/sec) by keeping stator reactive power equal to Zero level "Q_s = 0 (Var)"; to ensure only the exchange of the stator active power to the grid; means following the maximum active power point.

3- Mode 3 (Green color/ Fig. **3.36**): With MPPT strategy, in this case we propose a medium wind speed based on random form (Max wind speed = 13.5 m/sec) by keeping stator reactive power equal to Zero level "Q_s = 0 Var"; to ensure only the exchange of the stator active power to the grid; means following the maximum active power point.

● The second part focuses on the comparative simulation study between the *MPPT* strategy for conventional and proposed *IDPC* algorithms (*using PI, PID & MRAC respectively*) using step' and random wind speed, are described in detail in Figs. (**3.35** and **3.37**), respectively. This section is developed in order to illustrate the *Sub- and Super-synchronous* modes and the behavior of slip under generator speed variation.

● The third part deals with robustness tests using a comparative simulation study, for three modes (*with/without MPPT strategy*) for conventional *and* proposed *IDPC* algorithms (*using PI, PID & MRAC respectively*) is described in details in Figs. (**3.38-3.40**) respectively. This section is developed in order to verify the robustness of wind-system under parameter variation (*using three tests*[10]) in transient *and* steady states.

11.1. Mode 1 (Based on PI, PID and MRAC Without MPPT Strategy)

Fig. (**3.33**) shows the behavior of wind-system parameters of **Mode 1** under three topologies; the measured stator active and reactive powers (P_{s_meas} and Q_{s_meas}) and their references (P_{s_ref} and Q_{s_ref}) profiles are presented together in Fig. (**3.33a**) and are presented separately in Fig. (**3.33b,c**) respectively. The reference powers are indicated in Table **3.3**. The direct and quadrature components of currents and flux (I_{rd}, I_{rq} and Φ_{rd}, Φ_{rq}) are presented respectively in Fig. (**3.33e,g**), which present the inverse diagrams compared to reactive and active powers. The inverse case for stator direct and quadrature currents (I_{sd}, I_{sq}) which have the same diagrams of reactive and active powers, and they are presented in Fig. (**3.33d**). The power error is presented in Fig. (**3.33f**). The stator' and rotor currents; Is_abc and Ir_abc are shown in Fig. (**3.33h,i**) respectively, we remark the sinusoidal form of the three rotor and stator phases currents, have a good THD of stator currents (<= 5% respect the IEEE-519 Std) especially for PID and MRAC topologies. The power factor (PF) of the conventional and proposed control are presented in Fig. (**3.33j**), knowing that PF is the ratio of P to S (apparent power), had the top value when the reactive power equals to zero value; refer to these time interval in Figs. (**3.33a,b,c and j**): 0-0.2 (sec), 0.4-0.6 (sec), 0.8-1.0 and 1.2-1.4(sec).

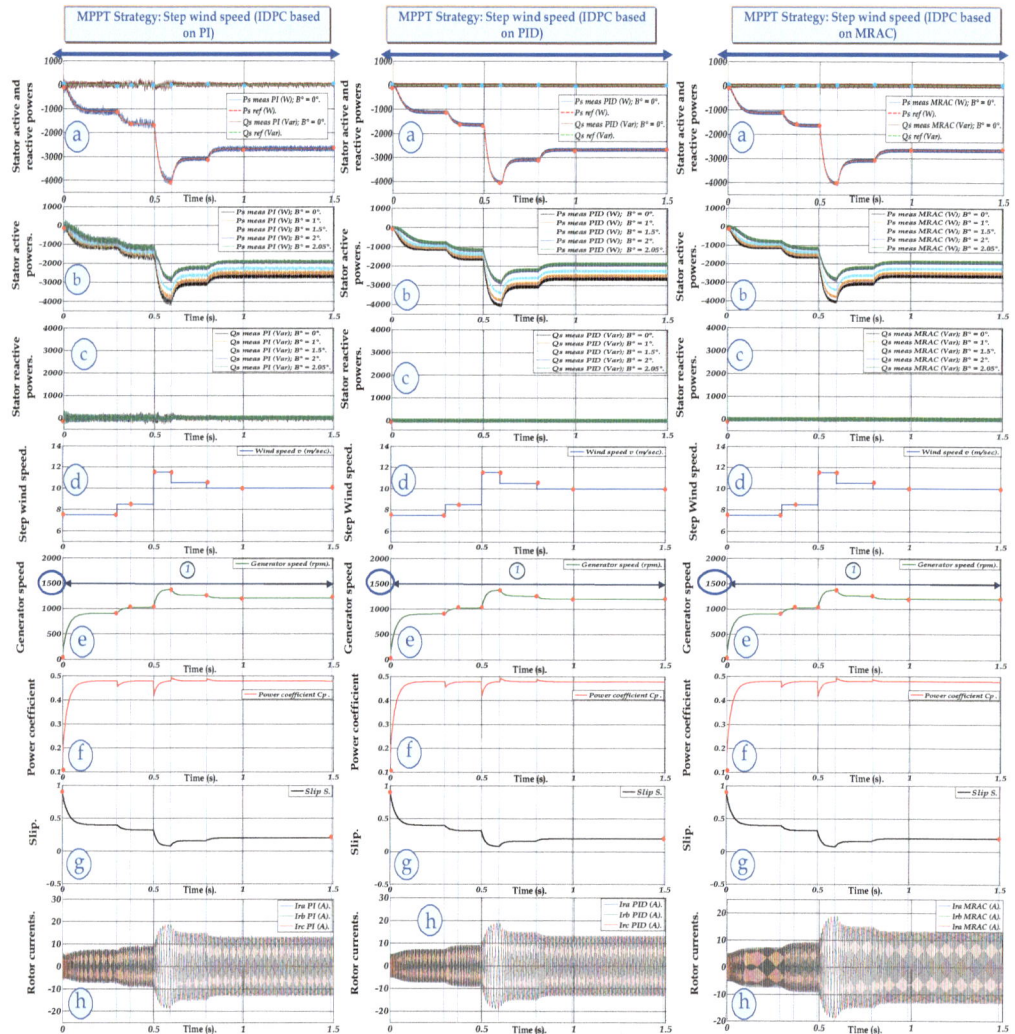

Fig. (3.35). Simulations results of two MPPT strategies based on **Mode-2** (using PI, PID & MRAC respectively); (**a**): stator active and reactive powers, (**b**): stator active power using different B° pitch angles, (**c**): stator reactive power using different pitch angles B°, (**d**): wind speed, (**e**): generator speed, (**f**): power coefficient, (**g**): slip, (**h**): rotor currents.

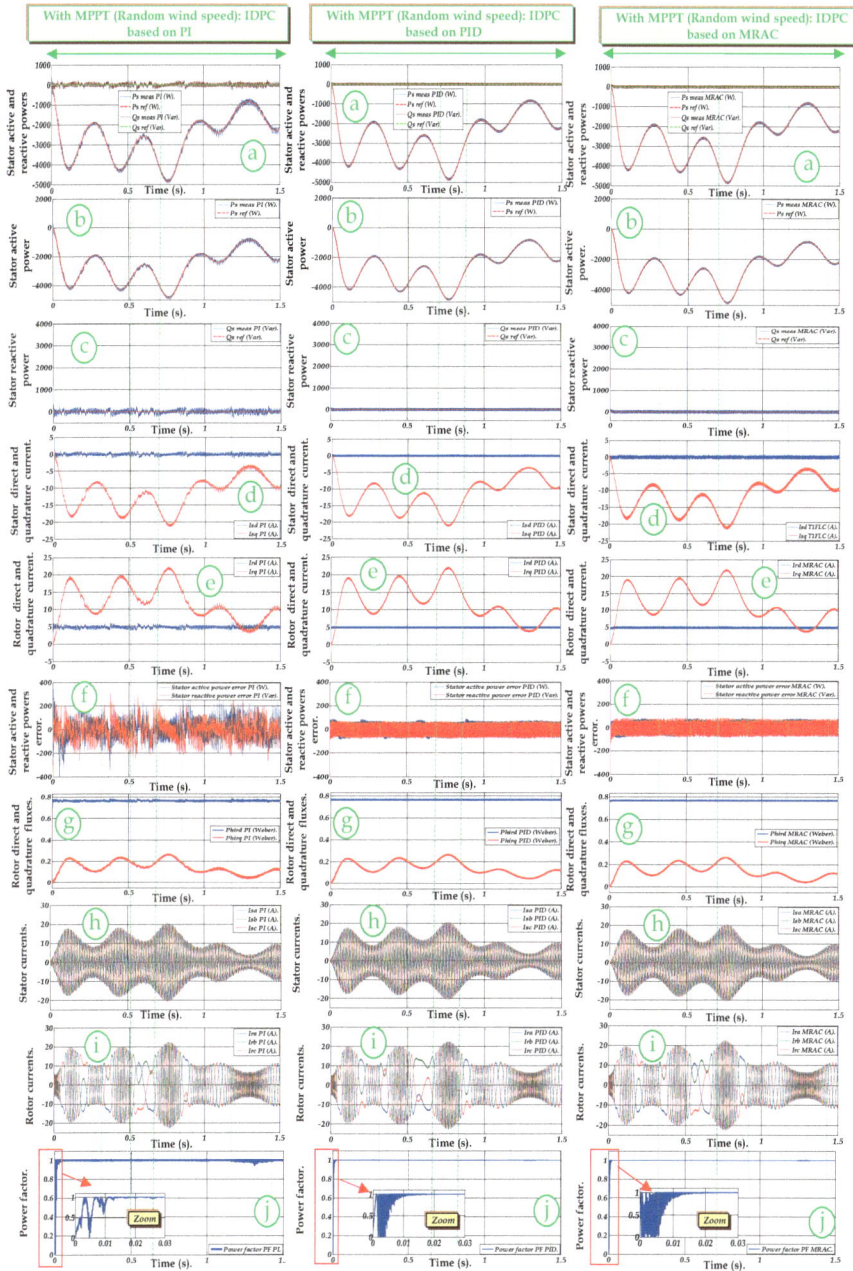

Fig. (3.36). Simulation results of **Mode-3** using PI, PID & MRAC; **(a)**: stator active and reactive powers, **(b)**: stator active power, **(c)**: stator reactive power, **(d)**: stator direct and quadrature currents, **(e)**: rotor direct and quadrature currents, **(f)**: stator active and reactive power error, **(g)**: rotor direct and quadrature fluxes, **(h)**: stator currents, **(i)**: rotor currents, **(j)**: power factor.

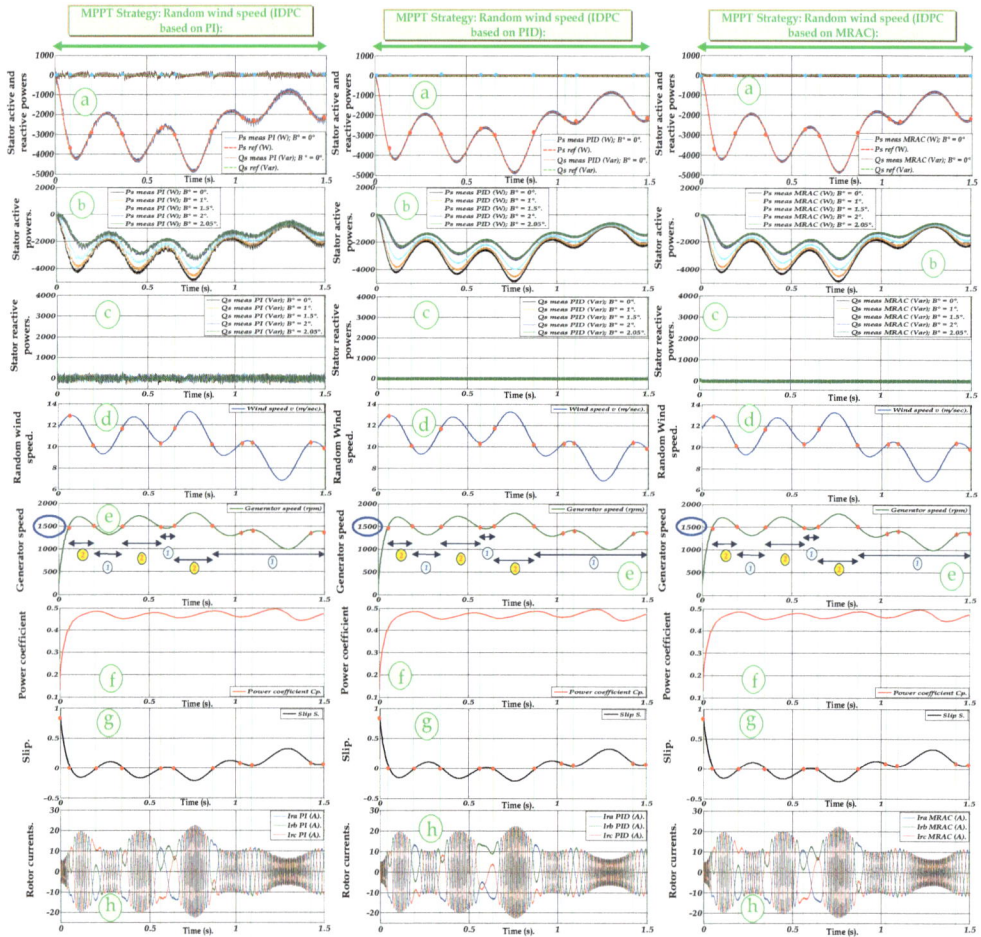

Fig. (3.37). Simulations results of two MPPT strategies based on **Mode-3** (using PI, PID & MRAC respectively); **(a)**: stator active and reactive powers, **(b)**: stator active power using different B° pitch angles, **(c)**: stator reactive power using different pitch angles B°, **(d)**: wind speed, **(e)**: generator speed, **(f)**: power coefficient, **(g)**: slip, **(h)**: rotor currents.

Table 3.3. The proposed profiles of the active and reactive power references.

Time (sec)	Stator Active Power (W)	Stator Reactive Power (Var)
[0 - 0.2]	-700.	0.
[0.2–0.4]	-1400.	-1400.
[0.4–0.6]	-700.	0.
[0.6–0.8]	-1400.	+1400.
[0.8–1.0]	-700.	0.
[1.0–1.2]	-1400.	-1400.
[1.2–1.4]	-700.	0.
[1.4–1.5]	-1400.	+1400.

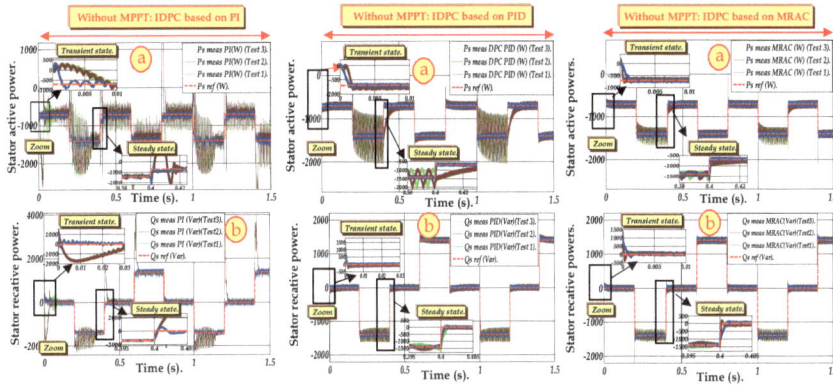

Fig. (3.38). Robustness tests of **Mode 1** for proposed control using PI, PID & MRAC respectively; (**a**): stator active powers, (**b**): stator reactive powers.

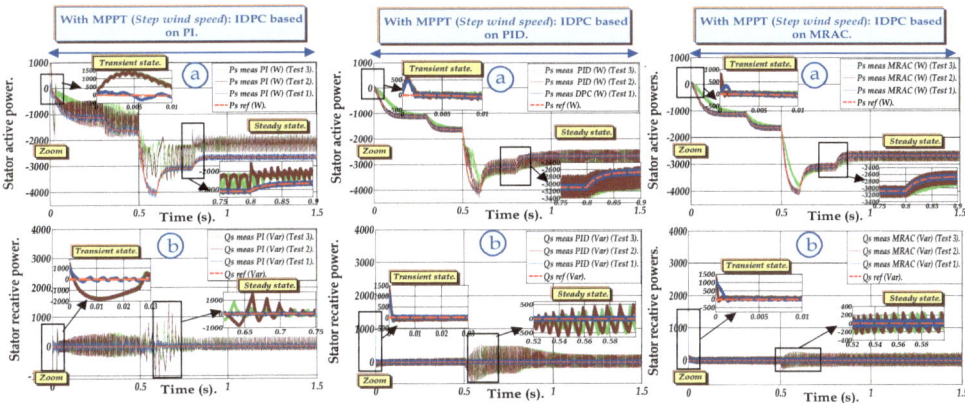

Fig. (3.39). Robustness tests of **Mode 2** for proposed control using PI, PID & MRAC respectively; (**a**): stator active powers, (**b**): stator reactive powers.

diagrams of reactive and active powers, and they are presented in Fig. (**3.33d**). The power error is presented in Fig. (**3.33f**). The stator' and rotor currents; I_{s_abc} and I_{r_abc} are shown in Fig. (**3.33h,i**) respectively, we remark the sinusoidal form of the three rotor and stator phases currents, have a good THD of stator currents (<= 5% respect the IEEE-519 Std) especially for PID and MRAC topologies. The power factor (PF) of the conventional and proposed control are presented in Fig. (**3.33j**), knowing that PF is the ratio of P to S (apparent power), had the top value when the reactive power equals to zero value; refer to these time interval in Figs. (**3.33a,b,c and j**): 0-0.2 (sec), 0.4-0.6 (sec), 0.8-1.0 and 1.2-1.4(sec).

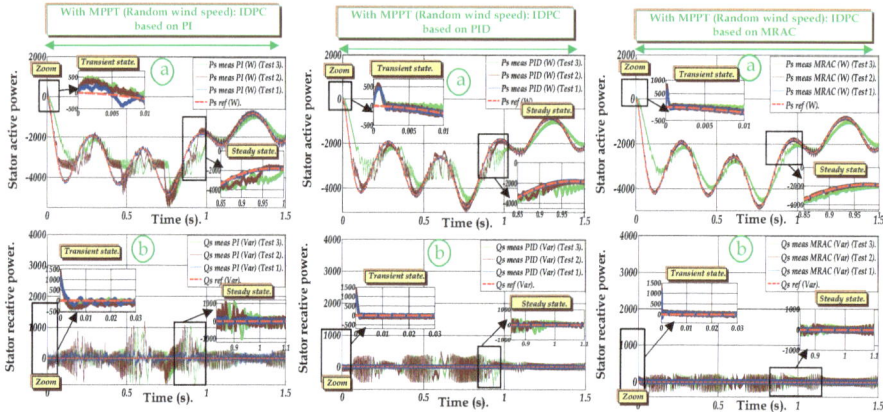

Fig. (3.40). Robustness tests of **Mode 3** for proposed control using PI, PID & MRAC respectively; (**a**): stator active powers, (**b**): stator reactive powers.

A- <u>IDPC based on PI:</u> (Fig.3.33) to the left side):

It is clear that the measured powers (active and reactive) have bad tracking power (big error) compared to their reference powers in transient and steady states (refer to Fig. (**3.33b,c**) to the left side), a remarkable (70%) overshoot is noted at 0.2, 0.4, 0.6, 0.8, 1.0, 1.2, 1.4 and 1.5 (sec) respectively in measured stator active' and reactive power (P_{s_meas} and Q_{s_meas}) -*because the PI controller is not robust in terms of overshoot especially if the step power' changement is big*-. We observe a big power error of active and reactive powers -200 (W_Var) $\leq \Delta P_s\_\Delta Q_s \leq$ +200 (W_Var). Table **3.4** depicts in details the obtained results of the conventional IDPC based on PI.

B- <u>IDPC based on PID:</u> (Fig. (3.33) to the middle side):

The measured powers (P_{s_meas} and Q_{s_meas}) have good tracking power (*refer to* Fig. (**3.33b,c**) *to the middle side*), a neglected overshoot (<5%) is noted in interval: 0.2 to 1.5 (sec) respectively in P_{s_meas} and Q_{s_meas}-*because the PID controller is robust in terms of overshoot especially if the step power' changement is big*-. We observe few power errors: -80 (W_Var) $\leq \Delta Ps\_\Delta Qs \leq$ +80 (W_Var). The obtained results of the proposed IDPC based on PID; are described in details in Table **3.4**.

C- <u>IDPC based on MRAC:</u> (Fig. (3.33) to the right side):

It is noted that the measured powers (P_{s_meas} and Q_{s_meas}) have good tracking power (*refer to* Fig. (**3.33b,c**) *to the right side*), a neglected ($\approx$ 5%) overshoot is noted is noted in interval: 0.2 to 1.5 (sec) respectively in P_{s_meas} and Q_{s_meas}-*because the MRAC controller is robust in terms of overshoot* -. We observe small power errors: $-80(\text{W\_Var}) \leq \Delta P_s, \Delta Q_s \leq +80(\text{W\_Var})$. The obtained results of the proposed IDPC based on MRAC; are shown in details in Table **3.4**.

Table 3.4. Recapitulation results for the proposed control using PI, PID & MRAC under mode 1.

	THD_Is_abc (%)	THD_Ir_abc (%)	Overshoot (%)	Response Time (Sec)	Power Error (W_Var)
PI	01.23%	50.74%	Remarquable ($\approx 70\%$).	$1 * 10^{-3}$.	+/- 200.
PID	0.73%	39.27%	Neglected ($< 5\%$).	$1.2 * 10^{-4}$.	+/- 80.
MRAC	0.77%	38.50%	Neglected ($\approx 5\%$).	$1.45 * 10^{-3}$.	+/- 80.

11.2. Mode 2 (Based on PI, PID and MRAC with MPPT Strategy- Step Wind Speed)

Fig. (**3.34**) shows the behavior of wind-system parameters of **Mode II** under three topologies; the reference stator active power (P_{s_ref}) (Fig. **3.34a**) is extracted from MPPT strategy (in this case, the wind speed will take step form); it takes the inverse diagram of wind speed. The stator reactive power (Q_{s_ref}) equals to 0 (Var), represents power factor unity. The measured stator active and reactive powers (P_{s_meas} and Q_{s_meas}) and their references (P_{s_ref} and Q_{s_ref}) profiles are presented together in Fig. (**3.34a**) and are presented separately in Fig. (**3.34b,c**) respectively. The direct and quadrature components of currents and flux (I_{rd}, I_{rq} and Φ_{rd}, Φ_{rq}) are presented respectively in Fig. (**3.34e,g**), which present the inverse diagrams compared to reactive and active powers. The inverse case for stator direct and quadrature currents (I_{sd}, I_{sq}) which have the same diagrams of reactive and active powers, and they are presented respectively in Fig. (**3.34d**). The power error is presented in Fig. (**3.34f**). The stator' and rotor currents; I_{s_abc} and I_{r_abc} are shown in Figs. (**3.34h,i**) respectively, we remark the sinusoidal form of the waveforms and have an excellent THD of stator currents ($<< 5\%$ respect the IEEE-519 Std) especially for PID and MRAC topologies. The power factor (PF) of the conventional and proposed control is presented in Fig. (**3.34j**), it reaches the top value when the reactive power equals to zero value, in this case the stator reactive power equals to zero value means the PF had taken always the unity, and this is the main aim of MPPT strategy-keeping at any time the unity power factor regardless wind speed variation.

A- <u>IDPC based on PI:</u> (Fig.3.34 to the left side):

It is noted that the poor power tracking of the measured powers compared to their reference powers in transient and steady states (refer to Fig. (**3.34b,c**)), a remarkable overshoot (40%) is noted at 0.6 (sec) in in measured stator active' and reactive power (P_{s_meas} and Q_{s_meas}) -*because the PI controller is not robust in terms* *of overshoot especially if the step power' changement is big-.* We observe a big power error of active and reactive powers -180 (W_Var) $\leq \Delta P_s\_\Delta Q_s \leq$ +180 (W_Var). Table **3.5** shows in details the obtained results of the conventional IDPC based on PI.

B- <u>IDPC based on PID:</u> (Fig.3.34 to the middle side):

It is clear that the measured powers (active and reactive) have good tracking power compared to their reference powers in transient and steady states (refer to Fig. **3.34b,c**)) to the left side), a neglected overshoot (< 3%) is noted at 0.6 (sec) in P_{s_meas} and Q_{s_meas}-*because the PID controller is robust in terms of overshoot especially if the step power' changement is big-.* We observe a low power error of active and reactive powers -80 (W_Var) $\leq \Delta P_s\_\Delta Q_s \leq$ +80 (W_Var). Table **3.5** depicts in details the obtained results of the proposed IDPC based on PID.

C- <u>IDPC based on MRAC:</u> (Fig.3.34 to the right side):

The measured powers (active and reactive) have good tracking power compared to their reference powers in transient and steady states, a neglected ($\approx$ 5%) overshoot is noted at 0.6 (sec) in P_{s_meas} and Q_{s_meas}-*because the MRAC controller is very robust in terms of overshoot especially if the wind power' changement is big-.* We observe a small power error of active and reactive powers -90 (W_Var) $\leq \Delta Ps\_\Delta Qs \leq$ +90 (W_Var). We remark the sinusoidal form of the waveforms of the stator' and rotor currents; I_{s_abc} and I_{r_abc} and excellent THD of stator currents will be injected into the grid (= 0.32% respect the IEEE-519 Std). Table **3.5** described in details the obtained results of the novel IDPC based on MRAC.

<u>**MPPT Strategy**</u>: Knowing that in this case (**Mode: 2**) the wind speed had taken the step form. Fig. (**3.35a**) presents the stator active and reactive powers and its references profiles for pitch angle equals to 0° (B°=0°) means the maximum power. Fig. (**3.35b**) illustrate the stator active power under different pitch angle (B°=0°,1°,1.5°,2° and 2.05°), it can be seen the inverse proportionality between the active powers and pitch angle and big undulations were noted. Fig. (**3.35c**) illustrate the stator reactive power under different pitch angle (B°=0°, 1°, 1.5°, 2° and 2.05°), it is clear that the reactive power doesn't maintain the zero level despite the Fig. (**3.35e**) presents the generator speed (<1500 (rpm) means the Sub-synchronous operation, noted by '**1**') and Fig. (**3.35e**) displays the generator speed (below and above the synchronous speed (1500 rpm) means the Sub-synchronous and Super-synchronous operations noted by '**1**' and '**2**' respectively). Fig. (**3.35f**) displays the power coefficient (Cp); this case Cp had maintained the maximum value despite of wind speed variation. Fig. (**3.35g**) displays the slip (S) behavior during the generator speed variation; Fig. (**3.35g**) illustrates the behavior of S under generator speed variation, in this case S varies between '**+1**' and '**0**'→ **0< S**

$\leq$+1 means the generator speed didn't reach the synchronous speed (always < 1500 (rpm)), Fig. (3.35g) represents the S behavior under generator speed variation, in this case S varies between '+1' and '-1' $\rightarrow$ -1< S $\leq$+1 means the generator speed varies below and above the synchronous speed (superior and inferior 1500 (rpm)), in the case when S reaches zero value means the mechanical speed N_r=1500 (rpm) this case called the synchronous mode (because S equals $(N_s$-$N_r)/N_s)$, for exp: in the case of mechanical speed N_r=1500 (rpm) and N_s =1500 =(60*f)/P=(60*50)/2=1500 (rpm) means S=(1500-1500)/1500 = 0) is the stable zone, please refer to Appendix. B, DFIG's parameters (*Table B.1*). Fig. (**3.35h**) displays the behavior of rotor currents under generator speed variation; Fig. (**3.35h**) illustrates the sinusoidal waveforms of rotor currents with remarkable ripples (*in this case the rotor currents did not change the sense because the Slip (S) did not reach the zero value*) and Fig. (**3.35h**) demonstrates the rotor currents behavior under generator speed variation, it can be seen the rotor currents had changed the sense in the case when S equals to zero value means when the generator speed varied near the synchronous speed.

Table 3.5. Recapitulation results for the proposed control using PI, PID & MRAC under mode 2.

	THD_Is_abc (%)	THD_Ir_abc (%)	Overshoot (%)	Respense Time (Sec)	Power Error (W_Var)
PI	0.78%	36.46%	Remarquable ($\approx$ 40%).	$1,3 * 10^{-3}$.	+/- 180.
PID	0.32%	5.89%	Neglected (< 3%).	$1,5 * 10^{-4}$.	+/- 80.
MRAC	0.32%	5.94%	Neglected ($\approx$ 5%).	$1.4 * 10^{-3}$.	+/- 90.

A- <u>IDPC based on PI:</u> (Fig.3.35 to the left side):

It is noted that the poor power tracking of the measured powers compared to their reference powers in transient and steady states (refer to Fig. **3.35a**), it can be seen remarkable ripples and undulation especially in steady states for the P_{s_meas} and Q_{s_meas} (Fig. **3.35a,b,c**), for this reason we conclude that the PI can not track the reference power with high performance.

B- <u>IDPC based on PID:</u> (Fig.3.35 to the middle side):

The good power tracking of the measured powers is noted compared to their reference powers in transient and steady states (refer to Fig. **3.35a,b,c**), prove that PID controller offer good tracking power especially if in wind speed severe variation compared to PI.

C- <u>IDPC based on MRAC:</u> (Fig. 3.35 to the right side):

It is noted that the good power tracking of the measured powers compared to their reference powers in transient and steady states (refer to Fig. **3.35a,b,c**), and prove also that the proposed controller called MRAC offer an excellent power tracking in severe conditions compared to the conventional one.

11.3. Mode 3 (Based on PI, PID and MRAC with MPPT Strategy- Random Wind Speed)

Fig. (**3.36**) shows the behavior of wind-system parameters of **Mode 3** under three topologies; the reference stator active power (P_{s_ref}) (Fig. **3.36a**) is extracted from MPPT strategy (in this case, the wind speed will take random form); it takes the inverse diagram of wind speed. The stator reactive power (Q_{s_ref}) equals to 0 (Var), represents power factor unity. The measured stator active and reactive powers (P_{s_meas} and Q_{s_meas}) and their references (P_{s_ref} and Q_{s_ref}) profiles are presented together in Fig. (**3.36a**) and are presented separately in Figs. (**3.34b,c**) respectively. The direct and quadrature components of currents and flux (I_{rd}, I_{rq} and Φ_{rd}, Φ_{rq}) are presented respectively in Fig. (**3.36e,g**), which present the inverse diagrams compared to reactive and active powers. The inverse case for stator direct and quadrature currents (I_{sd}, I_{sq}) which have the same diagrams of reactive and active powers, and they are presented respectively in Fig. (**3.36d**). The power error is presented in Fig. (**3.36f**). The stator' and rotor currents; I_{s_abc} and I_{r_abc} are shown in Figs. (**3.36h,i**) respectively, we remark the sinusoidal form of the waveforms and have an excellent THD of stator currents ($<< 5\%$ respect the IEEE-519 Std) especially for PID and MRAC topologies. The power factor (PF) of the conventional and proposed control is presented in Fig. (**3.36j**), it reaches the top value when the reactive power equals to zero value, in this case the stator reactive power equals to zero value means the PF had taken always the unity.

A- <u>IDPC based on PI:</u> (Fig.3.36 to the left side):

It is noted that the poor power tracking of the measured powers compared to their reference powers in transient and steady states (refer to Fig. **3.36b,c**), a big overshoot (40%) is noted at 0.5 (sec) and 0.8 (sec) in measured stator active' and reactive power (P_{s_meas} and Q_{s_meas}). We observe a big power error of active and reactive powers -250 (W_Var) $\leq \Delta P_s\_\Delta Q_s \leq +250$ (W_Var). Table **3.6** describes in details the obtained results of the conventional IDPC based on PI.

B- <u>IDPC based on PID:</u> (Fig.3.36 to the middle side):

It is noted that the poor power tracking of the measured powers compared to their reference powers in transient and steady states (refer to Fig. **3.36b,c**), a neglected overshoot (<3%) is noted at 0.5 (sec) and 0.8 (sec) in measured stator active' and reactive power (P_{s_meas} and Q_{s_meas}). We observe a lower power error of active and reactive powers -90 (W_Var) $\leq \Delta P_s, \Delta Q_s \leq$ +90 (W_Var). We remark the sinusoidal form of the waveforms and excellent THD of stator currents will be injected into the grid (= 0.25% and 0.26% for MRAC respect the IEEE-519 Std). Table **3.6** depicts in details the obtained results of the proposed IDPC based on PID.

C- <u>IDPC based on MRAC</u>: (Fig.3.36 to the right side):

It is noted that the poor power tracking of the measured powers compared to their reference powers in transient and steady states (refer to Fig. **3.36b,c**), a neglected *(≈ 5%)* overshoot is noted at 0.5 (sec) and 0.8 (sec) in measured stator active' and reactive power (P_{s_meas} & Q_{s_meas}) -*because the MRAC controller is ro-bust in terms of overshoot especially if the step power' changement is big-*. We observe a lower power error of active and reactive powers -90 (W_Var) $\leq \Delta P_s, \Delta Q_s \leq$ +90 (W_Var). Table **3.6** shows in details the obtained results of the novel IDPC based on MRAC.

<u>**MPPT Strategy**</u>: Knowing that in this case (**Mode: 3**); the wind speed had taken the random form. Fig. (**3.37a**) presents the stator active and reactive powers and its references profiles for pitch angle equals to 0° (B°=0°) means the maximum power. Fig. (**3.37b**) illustrate the stator active power under different pitch angle (B°=0°,1°,1.5°,2° and 2.05°), it can be seen the inverse proportionality between the active powers and pitch angle and big undulations were noted. Fig. (**3.37c**) illustrate the stator reactive power under different.

Table 3.6. Recapitulation results for the proposed control using PI, PID & MRAC under mode 3.

	THD_Is_abc (%)	THD_Ir_abc (%)	Overshoot (%)	Respense Time (Sec)	Power Error (W_Var)
PI	0.42%	36.46%	Remarquable (≈ 40%).	$1,3* 10^{-3}$.	+/- 250.
PID	0.25%	1.81%	Neglected (< 3%).	$1,6* 10^{-4}$.	+/- 90.
MRAC	0.26%	80.88%	Neglected (≈ 5%).	$1.3 * 10^{-3}$.	+/- 90.

Pitch angle (B°=0°, 1°, 1.5°, 2° and 2.05°), it is clear that the reactive power doesn't maintain the zero level despite the Fig. (**3.37e**) presents the generator speed and Fig. (**3.37e**) displays the generator speed (below and above the synchronous speed (1500 rpm) means the Sub-synchronous and Super-synchronous operations noted by '**1**' and '**2**' respectively). Fig. (**3.37f**) displays the power coefficient (Cp); this case Cp had maintained the maximum value despite of wind speed variation. Fig. (**3.37g**) displays the slip (S) behavior during the generator speed variation; Fig. (**3.37g**) illustrates the behavior of S under generator speed variation, in this case S varies between '**+1**' and '**0**'→ **0< S ≤+1** means the generator speed didn't reach the synchronous speed (always < 1500 (rpm)), Fig. (**3.37g**) represents the S behavior under generator speed variation, in this case S varies between '**+1**' and '**-1**'→ **-1< S ≤+1** means the generator speed varies below and above the synchronous speed (superior and inferior 1500 (rpm)), in the case when S reaches zero value means the mechanical speed N_r=1500 (rpm) this case called the synchronous mode (because S equals $(N_s-N_r)/N_s$), for exp: in the case of mechanical speed is the stable zone, please refer to Appendix **A**, DFIG's parameters (Table **A.1**). Fig. (**3.37h**) displays the behavior of rotor currents under generator speed variation; Fig. (**3.37h**) illustrates the sinusoidal waveforms of rotor currents with remarkable ripples (*in this case the rotor currents did not change the sense because the Slip (S) did not reach the zero value*) and Fig. (**3.37h**) demonstrates the rotor currents behavior under generator speed variation, it can be seen the rotor currents had changed the sense in the case when S equals to zero value means when the generator speed varied near the synchronous speed.

A- <u>IDPC based on PI:</u> (Fig.3.37 to the left side):

We note poor power tracking of the measured powers compared to their reference powers in transient and steady states (refer to Fig. **3.37a**), it can be seen remarkable ripples and undulation especially in steady states for the P_{s_meas} and Q_{s_meas} (Fig. **3.35a,b,c**), for this reason we conclude that the PI can't track the reference power with high performance.

B- <u>IDPC based on PID:</u> (Fig.3.37 to the middle side):

The good power tracking of the measured powers is noted compared to their reference powers in transient and steady states (refer to Fig. **3.37a,b,c**), prove that PID controller offer good tracking power especially if in wind speed severe variation compared to PI.

C- <u>IDPC based on MRAC:</u> (Fig.3.37 to the right side):

It is noted that the good power tracking of the measured powers compared to their reference powers in transient and steady states (refer to Fig. **3.37a,b,c**), and prove also that the proposed controller called MRAC offer an excellent power tracking in severe conditions compared to the conventional one.

11.4. Robustness Tests[12] for Mode 1, Mode 2 & Mode 3

Figs. (**3.38a,b**, **3.39a,b** and **3.40a,b**) illustrate the behavior of measured active and reactive powers and theirs references respectively under parameters variations; in transient and steady states.

A- Mode 1 (IDPC based on PI, PID & MRAC):

It can be noted big power in active and reactive power (topology based on PI/please refer to the left side of Fig. **3.38**) especially using the 2nd test and 3rd test (green color) with remarkable undulations especially in transient and steady states (please refer to zoom) and the value of power error reaches nearly ± 200 (W_Var) in Test-1, and nearly ±1000 (W_Var) in Test-2 and Test-3. A remarkable overshoot is noted under all robustness tests at: 0.2, 0.4, 0.6, 0.8, 1.0 1.2, 1.4 and 1.5 (sec).

For the second topology (based on PID/ refer to the middle side of Fig. **3.38**) a lower power error is noted for the Test-1 (blue color) reaches nearly ± 80 (W_Var), and nearly ±500 (W_Var) in Test-2 and Test-3. A remarkable overshoot is noted under all robustness tests especially at 0.2, 0.6 and at 1.0 (sec).

For the third topology (based on MRAC/ refer to the right side of Fig. **3.38**) also a small power error is noted for the first Test (blue color), and the value of power error reaches nearly ± 80 (W_Var) in Test-1, and nearly ±200 (W_Var) in Test-2 and Test-3. A few overshoot is noted under all robustness tests especially at 0.3 (sec) and at 1.0 (sec).

B- Mode 2 (IDPC based on PI, PID & MRAC):

Fig. (**3.39**)-(based on PI/please refer to the left side) display the behavior of stator active and reactive powers under MPPT strategy by maintaining the reactive power equals to zero value. In this case the active power had taken the inverse step profile of wind speed. Using robustness tests a remarkable undulations are noted (using tests: 2 and 3) especially at 0.6 (sec) which presents the rated power of DFIG (P=4 (kW)), on other hand and in the same time a remarkable power error is noted in stator reactive power (*which means that the PI controllers can't maintain the unity power factor "PF=1" under maximum wind power and parameters variation*), also a very big overshoot is noted in transient and steady states of active and reactive power.

For the second topology (based on PID/ refer to middle side of the Fig. **3.39**) the active power had taken the inverse step profile of wind speed. Using robustness tests a remarkable undulations are noted, on other hand and in the same time a remarkable power error is noted in stator reactive (*which means that the PID can't maintain the unity power factor under parameters variation*), also a remarkable overshoot is noted in transient and steady states of P_{s_meas} & Q_{s_meas}.

In the third topology (based on MRAC/ refer to right side of the Fig. **3.39**) a remarkable ripples are noted (using tests: 2 and 3) especially at 0.6 (sec) which presents the rated power of DFIG (P=4 (kW)), on other hand and in the same time a remarkable power error is noted in stator reactive power, also a few overshoot is noted in transient and steady states of active and reactive power (*which means that the MRAC can't maintain the unity power factor under parametersvariation*), also a lower overshoot is noted in transient and steady states of P_{s_meas} & $Q_{s\_meas.}$

C- <u>Mode 3 (IDPC based on PI, PID & MRAC):</u> (Fig.3.40 to the right side):

Fig. (**3.40a**)-(topology based on PI/refer to the left side of the Fig. **3.40**) display the behavior of stator active and reactive powers under MPPT strategy by maintaining the reactive power equals to zero value. In this case the active power had taken the inverse random profile of wind speed. Using robustness tests a remarkable undulations are noted (using tests: 2 and 3) especially at 0.75 (sec) and 0.8 (sec) which presents the over rated power of DFIG (P=4 (kW) and the measured active power maintain 4.6 (kW)), on other hand and in the same time a remarkable power error is noted in stator reactive power (*which means that the PI controllers can't maintain the unity power factor under maximum wind power*), also a very high overshoot is noted in transient and steady states of active and reactive power.

In the second topology (based on PID/refer to the middle side of the Fig. **3.40**) a remarkable undulations are noted (using tests: 2 and 3) which presents the over rated power of DFIG, on other hand and in the same time a remarkable power error is noted in stator reactive power (*which means that the PID controllers can't maintain the unity power factor under parameters changing*), also a big overshoot is noted in transient and steady states of active and reactive power (*please refer to the zoom*).

For the third topology (based on MRAC/refer to right side of the Fig. (**3.40**) acceptable ripples are noted (*using tests: 2 & 3*) and in the same time an acceptable power error and overshoot are shown in stator reactive power especially in steady states (*Please refer to the zoom*).

CONCLUSION

This chapter presents a detailed power control study (IDPC based on classical PI, PID & MRAC controllers) for DFIG-grid connection. In order to control independely DFIG's stator power; classical indirect power control has been combined with SVM to adjust active and reactive powers and rotor currents. MPPT strategy was proposed in order to extract the maximum wind-power despite unexpected wind speed variation (knowing that in this chapter we proposed two wind speed profiles: the step wind speed and the random wind speed) and to maintain power factor at unity (PF≈1). Several drawbacks appear in transient and steady states, in terms of tracking power and dynamic performances. In this context, improved IDPC algorithm was proposed. PID and MRAC controllers are proposed instead of PI to control stator powers and rotor currents respectively in order to overcome or mitigate the disadvantages of the conventional controllers. The simulation results have been developed *via MATLAB/ Simulink®* environment, show high dynamics response and improved wind-system performances using proposed algorithm compared to the conventional one. Using the robustness tests, the wind-performances are remarkably decreasing (in conventional and proposed algorithms respectively) which demonstrates the inability of tracking reference power during the parameter changement and sudden wind speed variation. In the next chapter, intelligent algorithms will be proposed in order to overcome or mitigate these drawbacks.

NOTES
[1]In this case the slip varies between $\pm 0.05 \rightarrow \pm 0.07$ (means the stable zone).
[2]Please refer to Appendix **A** (The gains of PI and PID controllers are illustrated in details in Table **A-4**).
[3]Schematic block of wind turbine is described in details in Fig. (**3.6**).
[4]Knowing that in reality Cp $=16/27 \approx 0.59$; (Betz limit) and this value is never achieved "Ref [18]" as shown in Fig. (**3.7**).
[5]Knowing that: T_{ij} (i=1, 2 'lines' and j=1, 2, 3 'columns').
[6]Knowing that: K_{ij} (i=1, 2 'are the lines' and j=1, 2, 3 'are the columns').
[7]In French literature, Sub-synchronous and Super-synchronous speeds are replaced by "Hypo-synchrone and Hypersynchrone".
[8]Knowing that the stable zone for generator' and motor' modes (please refer to Fig. **5.34**bold>) is within this range (band) 1400 (rpm); 1600 (rpm). From the stable zone it can be deduced that the slip S= (Ns-N)/Ns is equals to $\approx \pm (0.05$ to $0.07)$ means $\pm$ (5% to 7%).
[9]Please refer to Appendix **A** (PID's gain values are mentioned in Table **A.4**).
[10]Knowing that in this chapter the robustness tests are based on three tests as follows: [Test-1: without parameter changement $\rightarrow$ Blue color, Test-2: +100% of R_r and -25% of $(L_s, L_r$ and $L_m) \rightarrow$ Brown color and Test-3: +100% of (J and R_r), -25 % of $(L_r, L_s$ and $L_m)$ $\rightarrow$ Green color] respectively.
[11]Knowing that in this chapter the robustness tests are based on three tests as follows: [Test-1: without parameter changement $\rightarrow$ Blue color, Test-2: +100% of R_r and -25% of $(L_s, L_r$ and $L_m) \rightarrow$ Brown color and Test-3: +100% of (J and R_r), -25 % of $(L_r, L_s$ and $L_m)$ $\rightarrow$ Green color] respectively.

REFERENCES

[1]   G. Abad, J. Lopez, M.A. Rodriguez, L. Marroyo, and G. Iwanski, *"Doubly Fed Induction Machine Modeling And Control For Wind Energy Generation", IEEE Book.* John Wiley & Sons, 2011. [http://dx.doi.org/10.1002/9781118104965]

[2]   A. Tohidi, H. Hajieghrary, and M.A. Hsieh, "Adaptive Disturbance Rejection Control Scheme for DFIG-Based Wind Turbine: Theory and Experiments", *IEEE Trans. Ind. Appl.,* vol. 52, no. 3, pp. 2006-2015, 2016.
[http://dx.doi.org/10.1109/TIA.2016.2521354]

[3]   L. Li, H. Nian, L. Ding, and B. Zhou, *"Direct Power Control of DFIG System without Phase-Locked Loop under Unbalanced and Harmonically Distorted Voltage"*, *IEEE Trans.* Energ Conv, 2017.

[4]   H. Nian, and L. Li, "Direct Power Control of Doubly Fed Induction Generator without Phase-Locked Loop under Harmonically Distorted Voltage Conditions", *IEEE Trans. Power. Electron.,* 2017.

[5]   J. Mohammadi, and S. Vaez-zadeh, "A Combined Vector and Direct Power Control for DFIG-Based Wind Turbines", *IEEE Trans. Sustain. Energ,* vol. 5, no. 3, pp. 767-775, 2014.

[6]   R. Cardenas, R. Pena, S. Alepuz, and G. Asher, "Overview of Control Systems for the Operation of DFIGs in Wind Energy Applications", *IEEE Trans. Ind. Electron.,* vol. 60, no. 7, pp. 2776-2798, 2013.
[http://dx.doi.org/10.1109/TIE.2013.2243372]

[7]   E. Tremblay, S. Atayde, and A. Chandra, "Comparative Study of Control Strategies for the Doubly Fed Induction Generator in Wind Energy Conversion Systems: A DSP-Based Implementation Approach", *IEEE Trans. Sustain. Energ.,* vol. 2, no. 3, pp. 288-299, 2011.
[http://dx.doi.org/10.1109/TSTE.2011.2113381]

[8]   F. Amrane, and A. Chaiba, "A Novel Direct Power Control for grid-connected Doubly Fed Induction Generator based on Hybrid Artificial Intelligent Control with Space Vector Modulation", *Rev. Roum. Sci. Techn.– Électrotechn. et Énerg.,* vol. 61, no. 3, pp. 263-268, 2016.

[9]   F. Amrane, A. Chaiba, and A. Chebabhi, "Improvement Performances of Doubly Fed Induction Generator *via* MPPT Strategy using Model Reference Adaptive Control based on Direct Power Control with Space Vector Modulation", *J. Electric. Eng.,* vol. 16, no. 3, pp. 218-225, 2016. [JEE].

[10]   Y. Zou, M. E. Elbuluk, and Y. Sozer, "Stability Analysis of Maximum Power Point Tracking (MPPT) Method in Wind Power Systems", *IEEE Trans. Indust. Appl.,* vol. 49, no. 3, pp. 1129-1136, 2013.
[http://dx.doi.org/10.1109/TIA.2013.2251854]

[11]   S. Li, "Power Flow Modeling to Doubly-Fed Induction Generators (DFIGs) Under Power Regulation", *IEEE Trans. Power. Syst.,* vol. 28, no. 3, pp. 3292-3301, 2013.
[http://dx.doi.org/10.1109/TPWRS.2013.2251914]

[12]   B. Shen, B. Mwinyiwiwa, Y. Zhang, and B-T. Ooi, "Sensorless Maximum Power Point Tracking of Wind by DFIG Using Rotor Position Phase Lock Loop (PLL)", *IEEE Trans. Power Electron.,* vol. 24, no. 4, pp. 942-951, 2009.
[http://dx.doi.org/10.1109/TPEL.2008.2009938]

[13]   Z-S. Zhang, Y-Z. Sun, J. Lin, and G-J. Li, "Coordinated frequency regulation by doubly fed induction generator-based wind power plants", *IET Renew. Power Gener.,* vol. 6, no. 1, pp. 38-47, 2012.
[http://dx.doi.org/10.1049/iet-rpg.2010.0208]

[14]   N.K.S. Naidu, and B. Singh, "Experimental Implementation of a Doubly Fed Induction Generator Used for Voltage Regulation at a Remote Location", *IEEE Trans. Ind. Appl.,* vol. 52, no. 6, pp. 5065-5072, 2016.
[http://dx.doi.org/10.1109/TIA.2016.2600666]

[15]   G.D. Marques, and M.F. Iacchetti, "Stator frequency regulation in a field-oriented controlled DFIG connected to a DC link", *IEEE Trans. Ind. Electron.,* vol. 61, no. 11, pp. 5930-5939, 2014.
[http://dx.doi.org/10.1109/TIE.2014.2311403]

[16]   Y.A. Zorgani, Y. Koubaa, and M. Boussak, "MRAS state estimator for speed sensorless ISFOC induction motor drives with Luenberger load torque estimation", *ISA Trans.,* vol. 61, pp. 308-317, 2016.
[http://dx.doi.org/10.1016/j.isatra.2015.12.015] [PMID: 26775088]

[17]   S. Abdeddaim, A. Betka, S. Drid, and M. Becherif, "Implementation of MRAC controller of a DFIG based variable speed grid-connected wind turbine", *Energ. Conver. Manag.*, vol. 79, pp. 281-288, 2014.
[http://dx.doi.org/10.1016/j.enconman.2013.12.003]

[18]   R. Cárdenas, R. Peña, J. Clare, G. Asher, and J. Proboste, "MRAS observers for sensorless control of Doubly-Fed Induction Generators", *IEEE Trans. Power. Electron.*, vol. 23, no. 3, pp. 1075-1084, 2008.
[http://dx.doi.org/10.1049/cp:20080585]

[19]   L. Wang, and D.-N. Truong, "Stability Enhancement of DFIG-Based Offshore Wind Farm Fed to a Multi-Machine System Using a STATCOM", *IEEE Trans. Power. Syst.*, vol. 28, no. 3, pp. 2882-2889, 2013.
[http://dx.doi.org/10.1109/TPWRS.2013.2248173]

[20]   F. Amrane, and A. Chaiba, "Model Reference Adaptive Control for DFIG based on DPC with a Fixed Switching Frequency", In: *International Electrical Computer Engineering Conference, IECEC 23-25th May 2015 Setif-Algeria.*

[21]   N.K.S. Naidu, and B. Singh, "Grid-Interfaced DFIG-Based Variable Speed Wind Energy Conversion System With Power Smoothening", *IEEE Trans. Sustain. Energ.*, vol. 8, no. 1, pp. 51-58, 2017.
[http://dx.doi.org/10.1109/TSTE.2016.2582520]

[22]   A. Boyette, Contrôle-Commande d'un Générateur Asynchrone à Double Alimentation avec Système de Stockage pour la Production Eolienne, 2006.

[23]   F. Poitiers, Etude Et Commande De Generatrices Asynchrones pour L'utilisation de L'Energie Eolienne: -Machine asynchrone à cage autonome; -Machine asynchrone à double alimentation reliée au réseau, 2003.

[24]   L. Xiong, J. Wang, X. Mi, and M.W. Khan, "Fractional Order Sliding Mode Based Direct Power Control of Grid-Connected DFIG", *IEEE Trans. Power Syst.*, 2017.

[25]   D. Sun, X. Wang, H. Nian, and Z.Q. Zhu, "A Sliding-Mode Direct Power Control Strategy for DFIG under Both Balanced and Unbalanced Grid Conditions Using Extended Active Power", *IEEE Trans. Power Electron.*, vol. 33, no. 3, pp. 1313-1322, 2018.
[http://dx.doi.org/10.1109/TPEL.2017.2686980]

[26]   M. I. Martinez, and G. Tapia, "Sliding-Mode Control for DFIG Rotor- and Grid-Side Converters Under Unbalanced and Harmonically Distorted Grid Voltage", *IEEE Trans. Energ. Conv.*, vol. 27, no. 2, pp. 328-339, 2012.

[27]   Y. Bekakra, and D. Ben Attous, "DFIG sliding mode control fed by back-to-back PWM converter with DC-link voltage control for variable speed wind turbine", *Front. Energ.*, vol. 8, no. 3, pp. 345-354, 2014.
[http://dx.doi.org/10.1007/s11708-014-0330-x]

[28]   A. A. Tanvir, A. Merabet, and R. Beguenane, "Real-Time Control of Active and Reactive Power for Doubly Fed Induction Generator (DFIG)-Based Wind Energy Conversion System", *Energies,* vol. 8, 2015.
[http://dx.doi.org/10.3390/en80910389]

[29]   J. Mwaniki, H. Lin, and Z. Dai, "A Condensed Introduction to the Doubly Fed Induction Generator Wind Energy Conversion Systems", *Hindawi. J. Eng.*, vol. 8, pp. 1-18, 2017.
[http://dx.doi.org/10.1155/2017/2918281]

[30]   A. Susperregui, G. Tapia, I. Zubia, and J.X. Ostolaza, "Sliding-Mode control of Doubly-Fed Generator for Optimum Power Curve Tracking", *Electron. Lett.*, vol. 46, no. 6, pp. 126-127, 2010.
[http://dx.doi.org/10.1049/el.2010.0984]

[31]   K. Zhou, and D. Wang, "Relationship between Space-Vector Modulation and Three-Phase Carrier-Based PWM: A Comprehensive Analysis", *IEEE Trans. Ind. Electron.*, vol. 49, no. 1, pp. 186-196,

2002.
[http://dx.doi.org/10.1109/41.982262]

[32]　B. Vafakhah, *"Multilevel Space Vector PWM for Multilevel Coupled Inductor Inverters"*, University of Alberta, Ph.D. Thesis (English language), Canada, 2010.

[33]　G.A. Biain, *"Predictive Direct Control Techniques of The Doubly Fed Induction Machine for Wind Energy Generation Applications"*, PhD Thesis (English language), Mondragon Unibersitatea, Spain, 2008.

[34]　G.A. Biain, and G. Iwanski, "Chapter-10:"Properties and Control of a Doubly Fed Induction Machine" from the Book: "Power Electronics for Renewable Energy Systems, Transportation and Industrial Applications", In: *IEEE Press and John Wiley & Sons Ltd*, 2014.

[35]　F. Amrane, A. Chaiba, and S. Mekhilef, "High performances of grid-connected DFIG based on direct power control with fixed switching frequency *via* MPPT strategy using MRAC and Neuro-Fuzzy control", *J. Power. technol (JPT)*, vol. 96, no. 1, 2016.

CHAPTER 4

A Novel IDPC using Suitable Controllers (Robust and Intelligent Controllers)

Abstract: This chapter presents an improved Indirect power control (compared to the conventional one illustrated in chapter: 03) based on robust and suitable controllers (Robust and Intelligent controllers) to control the d-q axes currents (Ird and Irq) respectively. In order to overcome the speed/efficiency trade-off and divergence from peak power under fast variation of wind speed; three intelligent controllers (based on, T1-FLC, T2-FLC and NFC) are proposed to control the rotor direct and quadrature currents (Ird and Irq) instead of PID controllers, for grid-connected doubly fed induction generator (DFIG). The same wind-turbine (DFIG (4kW) and turbine (4.5 kW)) used in last chapter will be developed again in order to make a comparative study between the wind-system performance algorithms. The SVM strategy (to ensure the fixed switching frequency and to minimize the harmonics) is used in RSC for switching signals generation to control the inverter. In this chapter, mathematical model of each proposed controller is described in detail. The MPPT strategy is also developed in the three proposed algorithms in order to extract the maximum wind power by keeping the reactive power equal to zero value. The main aim of the proposed control is to improve the wind system performance despite the sudden wind speed variation and the DFIG's parameter variation in transient and steady states. The simulation results using the Matlab/Simulink environment (under three proposed modes and using robustness tests) show that the intelligent controller offered high power quality in spite of wind-speed variation have superior dynamic performance and are more robust during parameter variation.

Keywords: Neuro-Fuzzy Logic (NFC), Type-1 Fuzzy Logic Control (T1-FLC), Type-2 Fuzzy Logic Control (T2-FLC).

1. INTRODUCTION

In recent years, novel classes of fuzzy logic systems-well-known by type-2 fuzzy logic control (T2FLC) are valuable to incorporate the uncertainties is introduced [1]. Fuzzy logic controllers have the advantage of not needing a precise mathematical model, working with inaccurate inputs, managing non-linearity and being more robust than conventional PID controllers [2, 3]. Fuzzy Type 2 logic is capable of directing uncertainties because it can model them and reduce their effects compared to type 1 fuzzy logic control.

Fayssal Amrane & Azeddine Chaiba

Unfortunately, type-2 fuzzy sets are more challenging to understand and use than traditional type-1 fuzzy sets. Thus, their use is not prevalent yet. Even in the face of these problems, T2-FLC has found applications in the classification of coded video streams, co-channel interference removal from nonlinear time-varying communication channels, pre-processing radio-graphic images and transport scheduling [4, 5]. T2-FLCs found successful application in many engineering areas, demonstrating their aptitude to outperform T1-FLCs in the presence of dynamic uncertainties. The major difference between T1 and T2-FLCs is in the individual fuzzy sets model. T2 fuzzy sets employ membership degrees that are themselves fuzzy sets. This additional uncertainty dimension provides new degrees of freedom for modeling dynamic uncertainties. It noted that in recent years, the application of fuzzy logic is prolonged particularly in the wind energy conversion systems (WECS).

Furthermore fuzzy-neural techniques have been proposed as a robust control for electrical drives [6 - 8]. Neuro-fuzzy systems combine the advantages of neural networks and fuzzy logic systems. The principle purpose of using the ANFIS (*Adaptive Neuro-Fuzzy Inference Systems*) approach is to automatically provide the fuzzy system by using neural network (NN) methods [9, 10]. The main advantage of ANFIS architecture is modeling a highly Non-Linear system, as it combines the ability of fuzzy reasoning in manipulation the uncertainties and aptitude of an artificial neural network (ANN) in learning from processes.

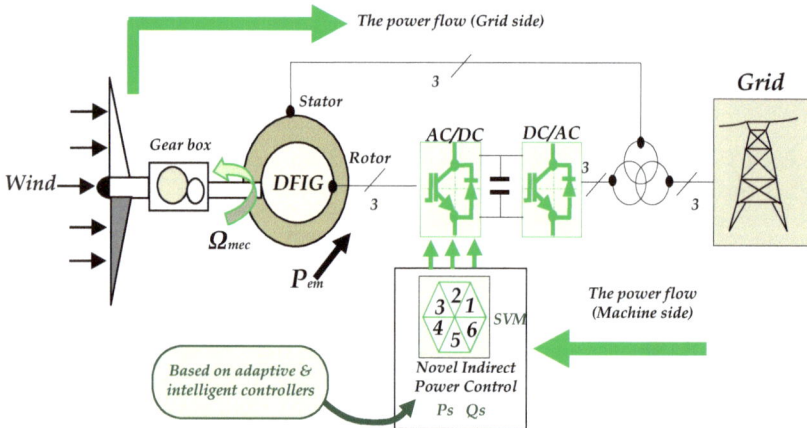

Fig. (4.1). Schematic diagram of wind-turbine DFIG based on novel indirect power control (IDPC).

A combination of the strengths of FLC and NNs creates systems able of controlling complex systems and adaptively learning to optimize control parameters [9, 10]. These attractive advantages justify the necessity of applying this kind of intelligent system for the wind energy conversion systems using

DFIG (Fig. **4.1**).

In this chapter the main contribution is the validation (*via MATLAB/Simulink®*) of the proposed IDPC based on intelligent controllers: T1-FLC, T2FLC and NFC instead the PID controller (as shown in Fig. **4.2**) in order to control rotor quadrature and direct currents (Ird and Irq) respectively under MPPT strategy. The robustness tests against parameters variation have been validated also using three MPPT modes in order to confirm that the drawbacks of the conventional IDPC are been treated. In order to more explain the advantages/disadvantages of all the proposed algorithms (*six proposed algorithms: chapter 03 + chapter 04*); a comparative study based on proposed controls will be proposed using several criteria.

2. DRAWBACKS AND PERFORMANCES LIMITATION OF CONVENTIONAL IDPC

The conventional algorithm presented in chapter 03 presents several drawbacks such as:

- An important overshoot is noted (more than + 50%).
- The coupling terms between the parameters of the both axis (d and q) has negative influence on the wind-system performances, especially in high wind power-wind generation (HWPG).
- The long response time -a visible delay of the measured value relative to that of the reference- order of $10e^{-2}$ (sec).
- A bad power tracking of the measured value relative to that of the reference especially if the profile is in the step form.
- Poor power/voltage quality which will be transmitted to the grid; a bad THD that exceeds IEEE standards (>> + 5%).
- A remarkable power error for conventional power control; sometimes exceeding at 25% of the rated power (± 1000 (W) for a rated power of 4 kW).
- The conventional regulators (PI regulators) depend on the DFIG's parameters.

In this context and to overcome these drawbacks; the adaptive and intelligent controllers are proposed instead the PID controllers, in order to improve power quality and wind-system performance, Fig. (**4.2**) present the adequate solution to ameliorate the power control under sudden wind speed variation and parameters changement. The mathematical model of each proposed controller is described in details in the next sections, then the simulation results are described in details (*simulation results for each controller topology under three modes "as described in chapter 03"*).

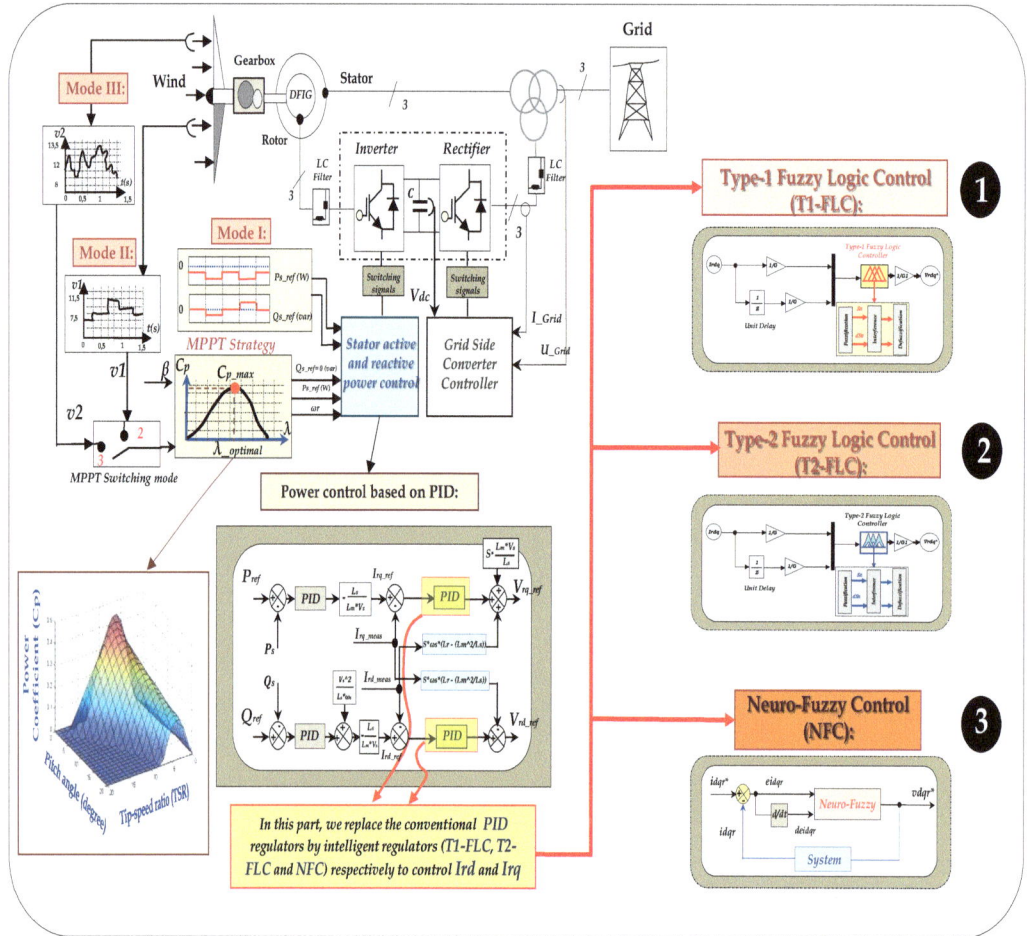

Fig. (4.2). Global wind-turbine system scheme based on three proposed controllers.

3. PROPOSED POWER CONTROL BASED ON TYPE-1 FUZZY LOGIC CONTROL (T1-FLC)

The *T1-FLC* consists of four stages; Fuzzification of inputs, derivation of rules based on knowledge, inference mechanism and defuzzification. In general, the aim of fuzzy logic system is to find a set of outputs for the given inputs in a nonlinear system, without using any mathematical model, but using linguistic rules [11].

3.1. Reasons for Choosing Fuzzy Logic

In general, the most powerful way of conveying information is by natural language. The conventional mathematical methods have not fully tapped this potential of language. The scientist [12] has said that the human thinking process is mainly based on conceptual patterns and mental images rather than on numerical quantities. So if the problem of assembling computers with the capability to work out complex issues has to be solved, the human thought process has to be modeled. The efficient way to do this is to use models that attempt to emulate the natural language; the creation of fuzzy logic has put this power to proper use. Many physical processes are not linear and especially to model them, a reasonable amount of approximation is necessary. For a simple system, the mathematical expressions give precise description of the system behavior.

Similarly, for more complicated systems with considerable amount of available data, model-free methods provide robust methods to reduce ambiguity and uncertainty in the system. But for complex systems with less numerical data, fuzzy reasoning furnishes a way to understand the system behavior by relying on approximate input-output approaches. The primary strength of fuzzy logic is that it makes use of linguistic variables rather than numerical variables to represent imprecise data [12].

3.2. Fuzzy Set Theory and Fuzzy Set Operations

In *1965*, *Zadeh* introduced the concept of fuzzy set theory. He states that, "*Much of the decision making in the real world takes place in an environment in which the goals, the constraints and the consequences of possible actions are not known precisely*". In recent years, fuzzy set theory applications have received increasing attention in designing intelligent controllers for complex industrial processes. We live in a world of marvelous complexity and variety where events never repeat exactly. Real world solutions are very often not crisp; but are vague, uncertain, and imprecise. Fuzzy logic provides us not only with meaningful and powerful representation for measurement of uncertainties but also with a meaningful representation of vague concepts in natural language. A fuzzy set can be defined mathematically by assigning to each possible individual, in the universe of discourse, a value representing its grade of membership in the fuzzy set. The basic three fuzzy set operations are union (U), intersection (∩) and complement (¯). The fuzzy logic method uses fuzzy equivalents of logical *AND*, *OR* and *NOT* operations to build up fuzzy logic rules [13]. In conventional set theory, *AND* is said to be the intersection of the sets and OR the union. The fuzzy operators based on values between zero and one, are sometimes said to be true generalizations of the Boolean operators.

3.3. Membership Functions

The membership functions play an important role in designing fuzzy systems [14]. The member-ship functions distinguish the fuzziness in a fuzzy set, whether the elements in the set are discrete or continuous in a graphical form for eventual use in mathematical formalism of fuzzy set theory. The shape of membership function describes the fuzziness in graphical form. The shape of membership functions is also important in the development of fuzzy system. The membership functions can be symmetrical or asymmetrical. A uniform representation of membership functions is desirable. The membership function defines how each point in the input space is mapped to a membership value in the interval [0, 1] [14].

3.4. Mamdani Fuzzy Inference Method

Generally, the starting point of a fuzzy system is the formation of a knowledge base consisting of *IF-THEN* rules. These rules are obtained from human experts based on their respective domain of knowledge and from observations that they made. Combining these rules into a single system is a natural step forward which allows us to obtain an output that attains the assigned goals. The *Mamdani's* fuzzy inference method is the most commonly used fuzzy methodology. *Mamdani's* method is the first control system built using fuzzy set theory. The fuzzy logic controller consists of four main parts. Among these, two of which perform transformations as shown in the Fig. (**4.3**) [18, 19].

The four parts are: **1:** Fuzzifier (Transformation 1), **2:** Knowledge Base, **3:** Inference Engine (Fuzzy Reasoning) and **4:** Defuzzifier (Transformation 2):

A- Fuzzifier

The fuzzifier performs measurement of the input variables (input signals, real variables), scale mapping and Fuzzification (Transformation 1). Thus, all the monitoring input signals are scaled and the measured signals (crisp input quantities which have numerical values) are transformed into fuzzy quantities by the process of Fuzzification. This transformation is performed by using membership functions. In a conventional fuzzy logic controller, the number of membership functions and the shapes of these are initially determined by the user. There are many different types of membership functions, piecewise linear or continuous. The commonly used membership functions are bell-shaped, sigmoid, Gaussian, triangular and trapezoidal. A membership function has a value between 0 and 1, and it indicates the degree of belongingness of a quantity to a fuzzy set.

B- Knowledge Base

The knowledge base consists of the data base and the linguistic control rule base. The data base provides the information which is used to define the linguistic control rules and the fuzzy data manipulation in fuzzy controller. The rule base contains a set of IF-THEN rules and these rules specify the control goal actions by means of a set of linguistic control rules. In other words, the rule base contains rules which would be provided by an expert.

C- Inference Engine

It is the kernel of a fuzzy logic controller and has the capability of both simulating human decision-making based on fuzzy concepts and of inferring fuzzy control actions by using fuzzy implication and fuzzy logic rules of inference. In other words, once all the monitored input variables are transformed into their respective linguistic variables, the inference engine evaluates the set of IF-THEN rules and thus the result is obtained which is again a linguistic value for the linguistic variable. Then this linguistic result has to be transformed into a crisp output value of the fuzzy logic control.

D- Defuzzifier

The second transformation is performed by the Defuzzifier which performs scale mapping as well as defuzzification. The Defuzzifier yields a non-fuzzy, crisp control action from the inferred fuzzy control action by using the consequent membership functions of the rules. There are many defuzzification techniques [13, 14]. They are center of gravity method, height method, mean of maxima method, first of maxima method, sum of maxima method, center of average method, etc.

Fuzzy logic is an approach to computing based on degrees of truth rather than the usual true or false, 1 or 0. Boolean deals with reasoning that are approximate rather than fixed and exact. Fuzzy logic is the methodology for the handling of inexact, imprecise, qualitative, fuzzy, verbal information in a systematic and rigorous way.

Fuzzy logic is used for the following reasons:

- Parameter variations that can be compensated with designer judgments.
- Processes that can be modeled linguistically but not mathematically.
- Setting with the aim to improve efficiency as a matter of operator skill and attention.
- When the system depends on operator skill and attention.

- Whenever fuzzy logic based controller can be used as an advisor to the human operator.
- Data intensive modeling.

The block diagram given in Fig. (**4.3**), fuzzy logic controller was simulated using triangular membership function and the centroid method was used for defuzzification.

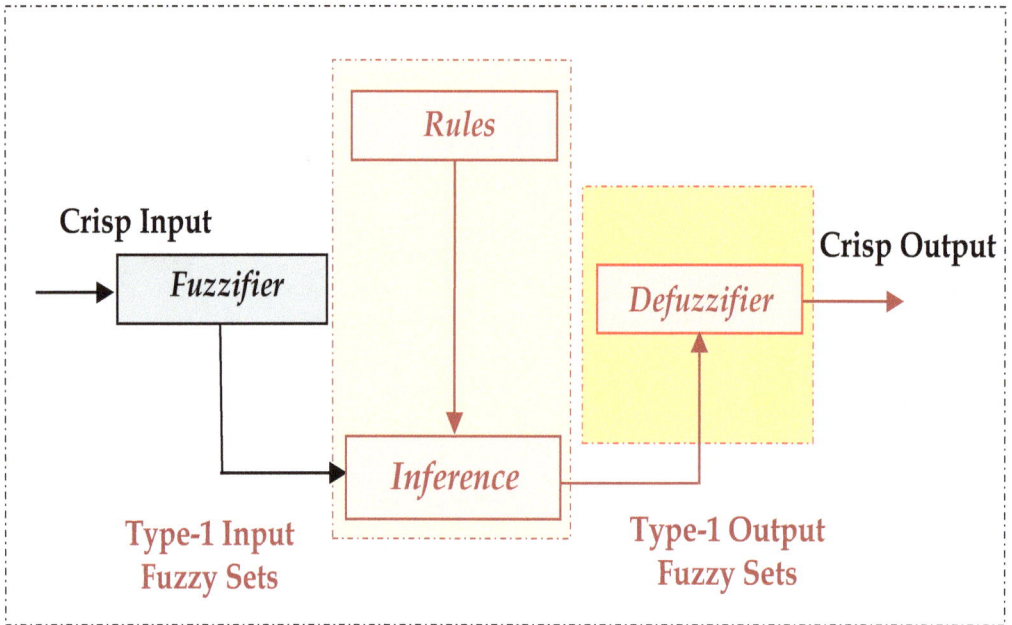

Fig. (4.3). Block diagram of Mamdani Type-1 fuzzy logic inference system.

3.5. MEMBERSHIP FUNCTIONS AND RULE BASE

Fuzziness in a fuzzy set is characterized by its membership functions. The membership functions convert the degree of fuzziness into the normalized interval *(0, 1)* where the boundary values *0* and *1* resemble membership degrees of crisp set members. The *T1-FLC* employed at *RSC* (*rotor side converter*) controller with seven (*07*) membership functions have been chosen for the inputs of: error rotor direct current (eI_{rd}), with its derivate (ed*Ird*) and the output is the (V_{rd}); knowing that the same member-ship function will be applied for the rotor quadrature current (*Irq*). The purpose of *T1-FLC* in *RSC* controller is mainly used to stabilize the measured rotor direct and quadrature current (I_{rd} and I_{rq}) at their references. The chosen membership functions are triangular form which are the most accepted and balanced choice in many applications [15].

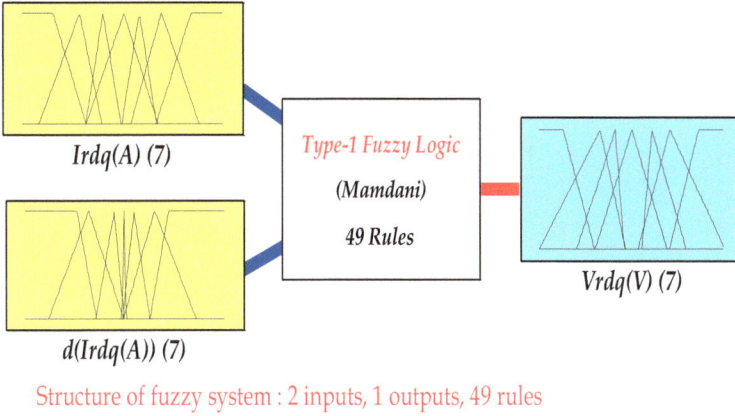

Structure of fuzzy system : 2 inputs, 1 outputs, 49 rules

Fig. (4.4). Global Memberships structure (2 inputs and 1 output).

The seven (07) inputs and output membership functions (Fig. **4.4**) are linguistically described as Negative Big-1 (*NB1*), Negative Medium-1 (*NM1*), Negative Small-1 (*NS1*), Zero-1 (*ZE1*), Positive Small-1 (*PS1*), Positive Medium-1 (*PM1*) and Positive Big-1 (*PB1*); knowing that "1" means the type- 1 fuzzy logic. Fig. (**4.5**) illustrates Schematic structure of *Type-1* fuzzy logic controller under *Matlab®/Simulink*. Fig. (**4.6**) depicts the simulation results of the membership functions using Matlab software; a and b: 2 inputs, c: 1 output and 4: the fuzzy logic surface).

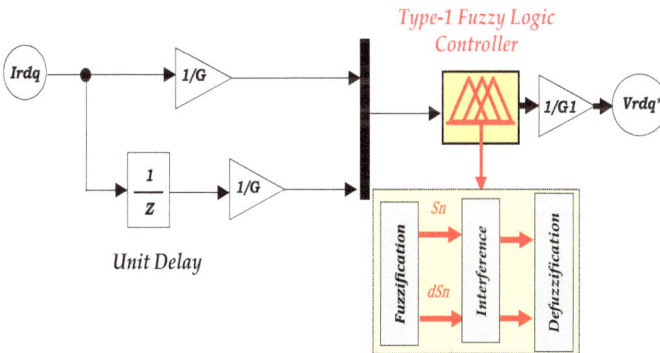

Fig. (4.5). Schematic structure of Type-1 fuzzy logic controller under Matlab®/Simulink.

The purpose of fuzzy logic controller is to make humanlike decisions by using the knowledge about controlling a target system. This is achieved by suitable fuzzy rules that constitute a fuzzy rule base. The fuzzy rules are formulated by means of *IF-THEN* rules.

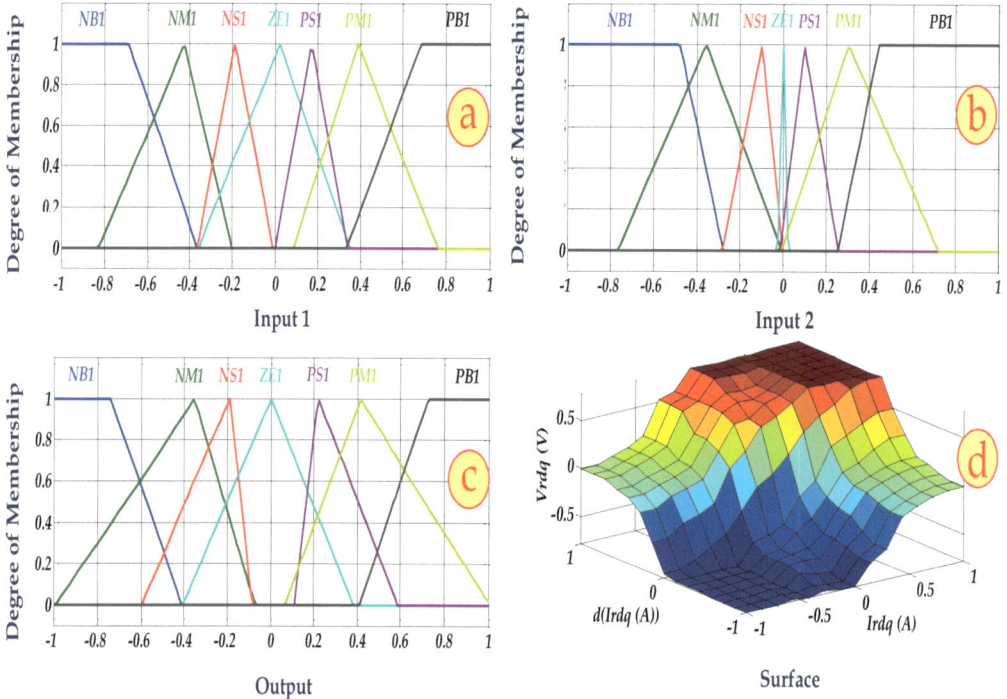

Fig. (4.6). Simulation results of the membership functions (a and b: 2 inputs, c: 1 output and d: the fuzzy logic surface).

Table. 4.1 Type-1 fuzzy logic inferences.

U1,2		dS1,2						
		NB1	**NM1**	**NS1**	**EZ1**	**PS1**	**PM1**	**PB1**
S1,2	**NB1**	NB1	NB1	NB1	NM1	NS1	NS1	**EZ1**
	NM1	NB1	NM1	NM1	NM1	NS1	**EZ1**	PS1
	NS1	NB1	NM1	NS1	NS1	**EZ1**	PS1	PM1
	EZ1	NB1	NM1	NS1	**EZ1**	PS1	PM1	PM1
	PS1	NM1	NS1	**EZ1**	PS1	PS1	PM1	PB1
	PM1	NS1	**EZ1**	PS1	PM1	PM1	PM1	PB1
	PB1	**EZ1**	PS1	PS1	PM1	PB1	PB1	PB1

The rule table which is shown in Table **4.1** contains 49 rules ($=7^2$). The structure of the fuzzy control rules for the two inputs and one output can be expressed as *if (eI_{rd} is PB1 and deI_{rd} is PB1) THEN output V_{rd} is PB1.*

Rule 1: if $S_{1,2}S1,2$ is NB1, and $dS_{1,2}$ is NB1 then $U_{1,2}dU1,2$ is NB1.
Rule 2: if $S_{1,2}S1,2$ is NM1, and $dS_{1,2}S1,2$ is NB1 then $U_{1,2}dU1,2$ is NB1.
Rule 3: if $S_{1,2}S1,2$ is NS1, and $dS_{1,2}S1,2$ is NG1 then $U_{1,2}dU1,2$ is NS1.

.

.

.

Rule 49: if $S_{1,2}S1,2$ is PB1, and $dS_{1,2}S1,2$ is PB1 then $U_{1,2}dU1,2$ is PB1.

The type-1 fuzzy rule base consists of a collection of linguistic rules[2] of the form [15 - 17].

In this case, the *T1-FLC* controller is proposed to control rotor *direct and quadrature* currents respectively (I_{rd} and I_{rq}) instead the *PID* controllers, in order to improve the dynamic responses against parameter variation. The simulation results of the proposed algorithms will presented in Figs. (**4.16**, **4.17** and **4.19**).

4. PROPOSED POWER CONTROL BASED ON TYPE-2 FUZZY LOGIC CONTROL (T2-FLC)

Type-2 fuzzy logic systems (*T2-FLS*) was introduced by *Zadeh* in *1975* as an extension of classical *Type-1* fuzzy logic systems (*T1-FLS*), are represented by fuzzy membership functions characterized by fuzzy sets in [0, 1] different a Type-1 fuzzy which have crisp membership functions [19].

Type-1 fuzzy logic controllers (FLCs) have established effective in dealing with complex nonlinear systems comprising uncertainties that are otherwise difficult to model or control. However, Type-1 FLCs have precise membership functions (MFs), *i.e.,* there is nothing uncertain in such MFs. Consequently, using Type-1 FLCs may cause degradation in the performance of some systems with noise and uncertainties. On the other side, the concept of Type-2 fuzzy sets is an extension of Type-1 fuzzy sets. The Type-2 fuzzy sets have grades of membership that are also fuzzy. The primary membership of Type-2 fuzzy sets can be any subset in [0, 1]. Moreover, corresponding to each primary membership, there is a secondary membership that is also in [0, 1], which defines the possibilities for the primary membership [19].

Both Type-1 fuzzy logic systems (FLSs) and Type-2 FLSs have the same four components: the fuzzifier, rule base, fuzzy inference engine, and output processor. Furthermore, unlike Type-1 FLSs, the output processor of Type-2 FLSs generates a Type-1 fuzzy set output using the type reducer or a crisp number using the Defuzzifier. A Type-2 FLS is also characterized by IF–THEN rules, but its antecedent or consequent sets are Type 2. Type-2 FLSs can be used when the circumstances are too uncertain to determine membership grades exactly, such as for imprecise or vague data. Thus, the Type-2 FLSs have supplanted conventional technologies in many applications, especially in complex nonlinear systems. However, in general, Type-2 FLSs are computationally intensive due to the complexity of reducing Type-2 FLSs to Type-1 FLSs [19]. To simplify the computation, the secondary membership grade can be set to one, in which case the Type-2 FLSs become interval Type-2 FLSs. Currently, the most widely used Type-2 fuzzy sets in Type-2 FLCs are interval Type-2 fuzzy sets.

4.1. Overview of Type-2 Fuzzy Logic Controller Toolbox

Fig. (**4.7**) shows the FIS Editor by using *Type-1* fuzzy logic control (A) and Type-2 fuzzy logic control (B) respectively *via MATLAB/Simulink® R2009a*. Fig. (**4.8**) depicts Type-2 fuzzy logic toolbox interface.

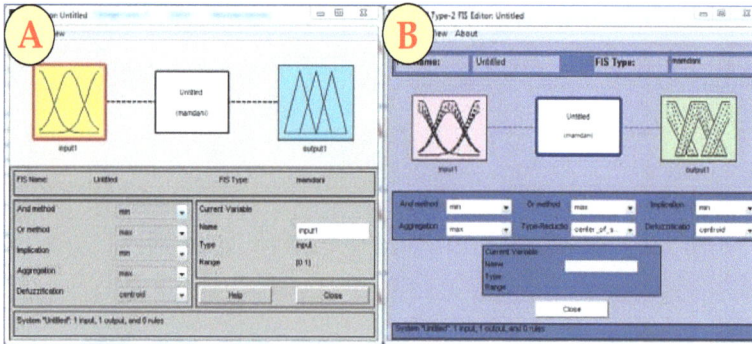

Fig. (4.7). FIS editor A: type-1 fuzzy logic toolbox, B: type-2 fuzzy logic toolbox.

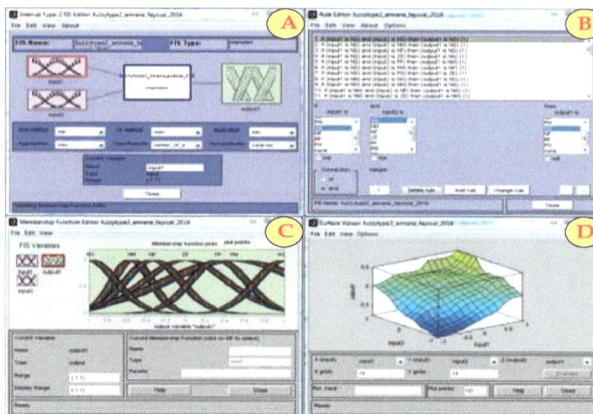

Fig. (4.8). Type-2 fuzzy logic interface; A: Membership function editor of Type-2 fuzzy logic toolbox, B: Rule editor of type-2 fuzzy logic toolbox, C: 2 Inputs-1output Membership function editor of type-2 fuzzy logic Toolbox, D: Surface viewer of type-2 fuzzy logic toolbox.

4.2. Design of Type-2 Fuzzy Logic Controller

As described in **4.1**, the Type-2 fuzzy logic control (*T2-FLC*) have the same principle as *T1-FLC*, is also composed from four parts are: **1:** Fuzzifier (*Transformation 1*), **2:** Knowledge Base, **3:** Inference Engine (*Fuzzy Reasoning*) and **4:** Defuzzifier (*Transformation 2*). But the structure is different than *T1-FLC*.

The structure of a *type-2* Fuzzy Logic System (*FLS*) is shown in Fig. (**4.9**). It is actually very similar to the structure of an ordinary *type-1 FLS*. It is assumed in this chapter that the reader is familiar with *type-1 FLSs* and thus, in this section, only the similarities and differences between *type-2* and *type-1 FLSs* are underlined.

The fuzzifier shown in Fig. (**4.9**), as in a *type-1 FLS*, maps the crisp input into a fuzzy set. This fuzzy set can be a *type-1*, *type-2* or a singleton fuzzy set. In singleton Fuzzification, the input set has only a single point of nonzero membership. The singleton fuzzifier is the most widely used fuzzifier due to its simplicity and lower computational requirements.

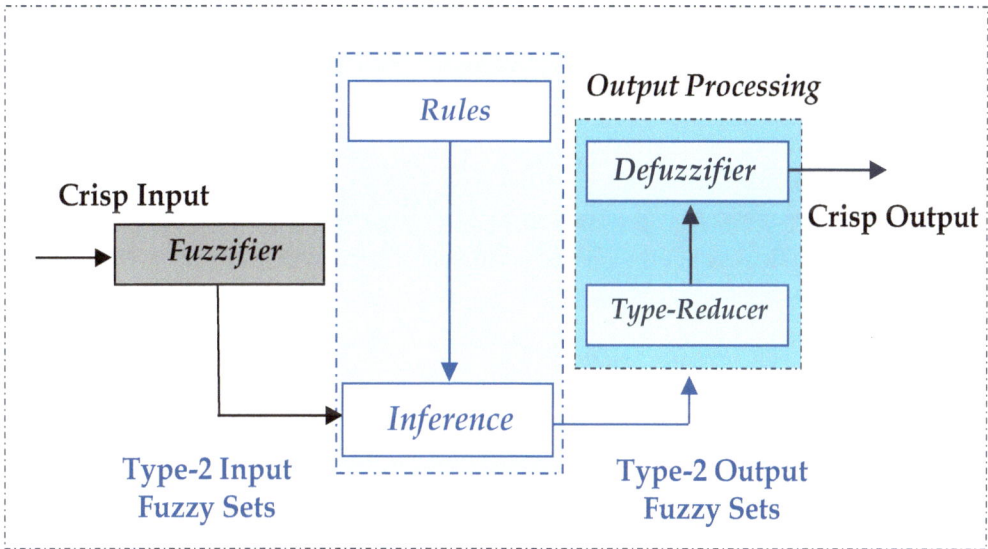

Fig. (4.9). Block diagram of Mamdani Type-2 fuzzy logic inference system

The shaded region in Fig. (**4.10a**) is the *FOU* for a *type-2* fuzzy set. The primary memberships: J_{x1} and J_{x2} and their associated secondary membership functions and are shown at the points x_1 and x_2. The upper and lower membership functions, and, are also shown in Fig. (**4.10a**). The secondary membership functions, which are interval sets, are shown in Fig. (**4.10b**) [1].

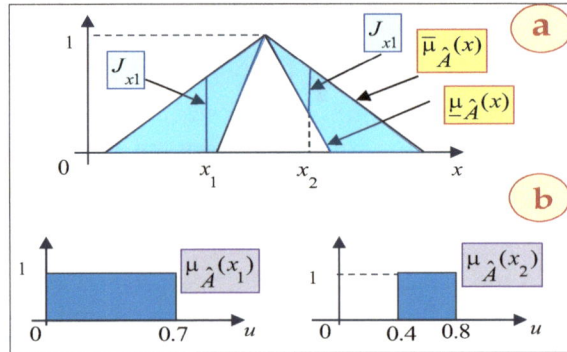

Fig. (4.10). -(a) The FOU for a type-2 fuzzy set, (b) The secondary membership functions.

The third dimension of *type-2* fuzzy sets decides secondary membership function and the *FOU* decides the range of uncertainty, together provides additional degree of freedom in the design to compensate various uncertainties. Hence, wind energy systems, being highly uncertain, can utilize the special features of *type-2 FLSs* to improve its operational efficiency in the grid interaction [20]. The type2 fuzzy controller utilized in this work has two inputs and one output.

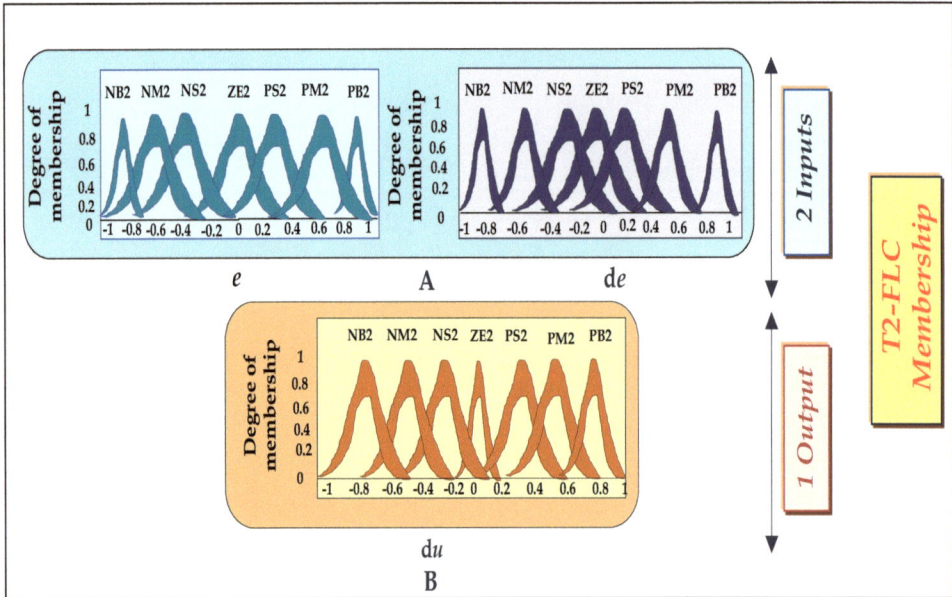

Fig. (4.11). Membership functions (A: inputs and B: output).

The membership functions are defined in Fig. (**4.11 a and b**). The inferences of *T2-FLC* can be made in a more explain as shown in Table **4.2** [21, 22]. The purpose of T2-FLC is to make humanlike decisions by using the knowledge about controlling a target system. This is achieved by suitable fuzzy rules that constitute a fuzzy rule base. The fuzzy rules are formulated by means of *IF-THEN* rules. The rule table which is shown in Table **3.2** contains 49 rules (=7^2). The structure of the fuzzy control rules for the two inputs and one output can be expressed as *if (eI_{rd} is PB2 and deI_{rd} is PB2) THEN output V_{rd} is PB2.*

The *type-2* fuzzy rule base consists of a collection of linguistic rules[1] of the form [22, 23]. *Knowing that*, the schematic structure of *type-2* fuzzy logic controller under *Matlab/Simulink® R2009a (in order to control I_{rd} and I_{rq})* is illustrated in details in Fig. (**4.12 a and b**).

In this case, the *T2-FLC* controller is proposed to control rotor direct and quadrature currents respectively (I_{rd} and I_{rq}) instead the PID controllers, in order to improve the dynamic responses against parameter variation. The simulation results of the proposed algorithms will presented in Figs.-(**4.16**, **4.17** and **4.19**).

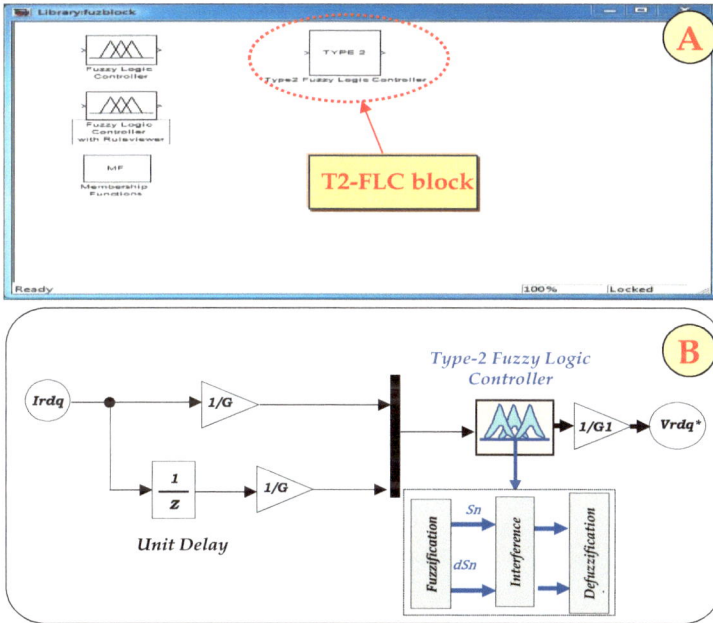

Fig. (4.12). (A and B) Schematic structure of Type-2 fuzzy logic controller under Matlab®/Simulink.

Rule 1: if $S_{1,2}$S1,2 is NB2, and $dS_{1,2}$is NB2 then $U_{1,2}$dU1,2 is NB2.
Rule 2: if $S_{1,2}$S1,2 is NM2, and $dS_{1,2}$S1,2is NB2 then $U_{1,2}$dU1,2 is NB2.
Rule 3: if $S_{1,2}$S1,2 is NS2, and $dS_{1,2}$S1,2 is NG2 then $U_{1,2}$dU1,2 is NS2.

.
.
.

Rule 49: if $S_{1,2}$S1,2 is PB2, and $dS_{1,2}$S1,2 is PB2 then $U_{1,2}$dU1,2 is PB2.

Table. 4.2. Type-2 fuzzy logic inferences [24].

U1,2		*dS1,2*						
		NB2	*NM2*	*NS2*	*EZ2*	*PS2*	*PM2*	*PB2*
S1,2	*NB2*	*NB2*	*NB2*	*NB2*	*NM2*	*NS2*	*NS2*	*EZ2*
	NM2	*NB2*	*NM2*	*NM2*	*NM2*	*NS2*	*EZ2*	*PS2*
	NS2	*NB2*	*NM2*	*NS2*	*NS2*	*EZ2*	*PS2*	*PM2*
	EZ2	*NB2*	*NM2*	*NS2*	*EZ2*	*PS2*	*PM2*	*PM2*
	PS2	*NM2*	*NS2*	*EZ2*	*PS2*	*PS2*	*PM2*	*PB2*
	PM2	*NS2*	*EZ2*	*PS2*	*PM2*	*PM2*	*PM2*	*PB2*
	PB2	*EZ2*	*PS2*	*PS2*	*PM2*	*PB2*	*PB2*	*PB2*

5. PROPOSED POWER CONTROL BASED ON NEURO-FUZZY CONTROL (*NFC*)

The block diagram of the neuro-fuzzy controller (*NFC*) system is shown in Fig. (**4.13**). The *NFC* controller is composed of an on-line learning algorithm with a neuro-fuzzy network. The neuro-fuzzy network is trained using an on-line learning algorithm. The *NFC* has two inputs, the rotor current error and the derivative of rotor current error . The output is rotor direct voltage . For the *NFC* of rotor current is similar with controller [25].

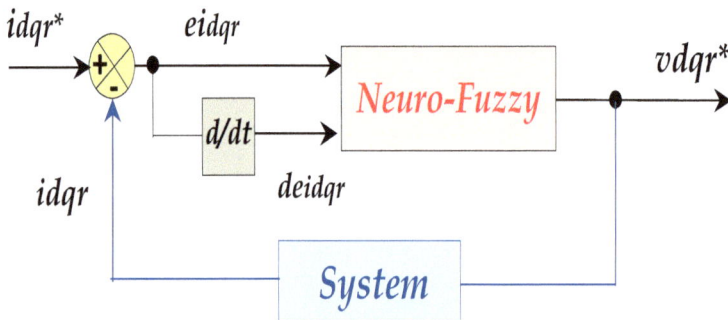

Fig. (4.13). Block diagram of the neuro-fuzzy controller under Matlab/Simulink®.

For the NFC, a four layer NN as shown in Fig. (**4.16**) is used. Layers I–IV represents the inputs of the network, the membership functions, the fuzzy rule base and the outputs of the network, respectively [25]. Knowing that: the parameters of the proposed NFC are illustrated in details in Table **4.3**. The training error is depicted in Fig. (**4.14**), which equals to 0.05; presents very good precision).

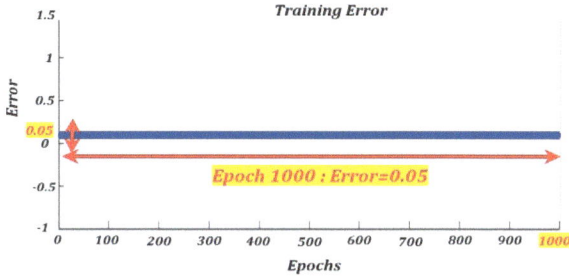

Fig. (4.14). Simulation results of training error (0.05) using 1000 epochs.

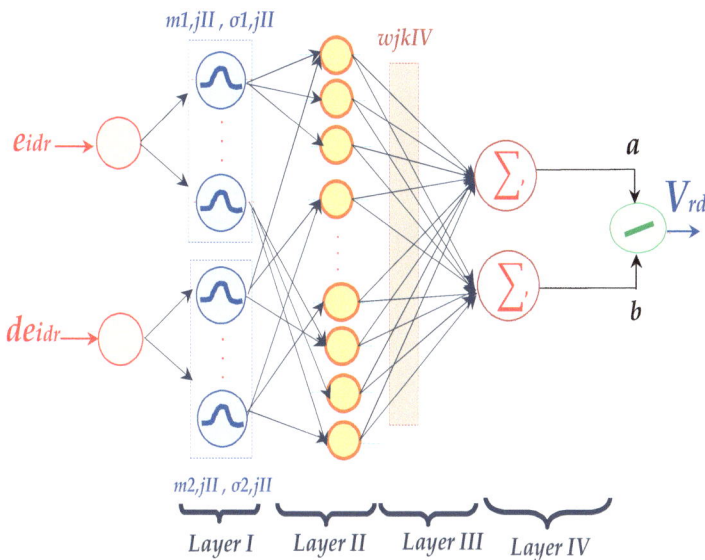

Fig. (4.15). Schematic diagram of the neuro-fuzzy network.

5.1. Layer I: Input layer

Inputs and outputs of nodes in this layer are represented as:

Table. 4.3 The parameters of the proposed Neuro-fuzzy controller.

Type:	Takagi Sugeno
Number of iteration:	500.
Error tolerance:	$5*10^{-3}$.
Epochs:	1000.
Number of membership function:	7.
Number of hidden layer neuron:	14.

$$net_1^I = e_{idr}(t), y_1^I = f_1^I(net_1^I) = net_1^I = e_{idr}(t) \tag{4.1}$$

$$net_2^I = \dot{e}_{idr}(t), y_2^I = f_2^I(net_2^I) = net_2^I = \dot{e}_{idr}(t) \tag{4.2}$$

Where e_{idr} and e_{idr} are inputs Y_1^I and Y_2^I are outputs of the input layer. In this layer, the weights are unity and fixed.

5.2. Layer II: membership layer

In this layer, each node performs a fuzzy set and the Gaussian function is adopted as a membership function.

$$net_{1,j}^{II} = \frac{-\left(x_{1,j}^{II} - m_{1,j}^{II}\right)^2}{\left(\sigma_{1,j}^{II}\right)^2}, y_{1,j}^{II} = f_{1,j}^{II}(net_{1,j}^{II}) = exp(net_{1,j}^{II}) \tag{4.3}$$

$$net_{2,k}^{II} = \frac{-\left(x_{2,k}^{II} - m_{2,k}^{II}\right)^2}{\left(\sigma_{2,k}^{II}\right)^2}, y_{2,k}^{II} = f_{2,k}^{II}(net_{2,k}^{II}) = exp(net_{2,k}^{II}) \tag{4.4}$$

Where: $m_{1,j}^{II}$, $m_{2,k}^{II}$ and $\sigma_{1,j}^{II}$, $\sigma_{2,k}^{II}$ are respectively, the mean and the standard deviation of the Gaussian function. There are $j + k$ nodes in this layer.

5.3. Layer III: rule layer

This layer includes the rule base used in the fuzzy logic control (*FLC*). Each node in this layer multiplies the input signals and outputs the result of product [6].

$$net_{j,k}^{III} = \left(x_{1,j}^{III} * x_{2,k}^{III}\right), y_{jk}^{III} = f_{jk}^{III}(net_{jk}^{III}) = net_{jk}^{III} \tag{4.5}$$

Where the values of link weights between the membership layer and rule base layer are unity.

5.4. Layer IV: output layer

This layer represents the inference and defuzzification used in the *FLC*. For defuzzification, the center of area method is used; therefore the following form can be obtained:

$$a = \sum_j \sum_k w_{jk}^{IV} y_{jk}^{III}, b = \sum_j \sum_k y_{jk}^{III} \tag{4.6}$$

$$net_0^{IV} = \frac{a}{b}, y_0^{IV} = f_0^{IV}(net_0^{IV}) = \frac{a}{b} \tag{4.7}$$

Where y^{III}_{jk} is the output of the rule layer; a and b are the numerator and the denominator of the function used in the center of area method, w^{IV}_{jk} is the center of the output membership functions used in the FLC, respectively.

The aim of the learning algorithm is to adjust the weights of , w^{IV}_{jk}, $m^{II}_{1,j}$, $m^{II}_{2,k}$, $\sigma^{II}_{1,j}$, $\sigma^{II}_{2,k}$. The on-line learning algorithm is a gradient descent search algorithm in the space of network parameters.

The error expression for the input of *Layer IV*:

$$\delta^{IV}_0 = \frac{\dfrac{-\partial e_{idr}(t)\dot{e}_{idr}(t)}{\partial y^{IV}_0} * \partial y^{IV}_0}{\partial net^{IV}_0} = \mu_5 e_{idr} \tag{4.8}$$

Where μ_5 is the learning-rate for w^{IV}_{jk} and it can be shown in following equation.

Therefore, the changing of w^{IV}_{jk} is written as:

$$\Delta w^{IV}_{jk} = \frac{\dfrac{-\partial e_{idr}(t)\dot{e}_{idr}(t)}{\partial net^{IV}_0} * \partial net^{IV}_0}{\dfrac{\partial a}{\partial w^{IV}_{jk}}} * \partial a = \frac{1}{b}\delta^{IV}_0 y^{III}_{jk} \tag{4.9}$$

Since the weights in the rule layer are unified, only the approximated error term needs to be calculated and propagated by the following equation:

$$\delta^{III}_{jk} = \frac{\dfrac{-\partial e_{idr}(t)\dot{e}_{idr}(t)}{\partial net^{IV}_0} * \partial net^{IV}_0}{\dfrac{\partial y^{III}_{1,j}}{\partial net^{III}_{jk}}} * \partial y^{III}_{1,j} = \frac{1}{b}\delta^{IV}_0\left(w^{IV}_{jk} - y^{IV}_0\right) \tag{4.10}$$

The error received from *Layer III* is computed as:

$$\delta^{II}_{1k} = \sum_k \left[\frac{\dfrac{\left(\dfrac{-\partial e_{idr}(t)\dot{e}_{idr}(t)}{\partial net^{III}_{jk}}\right) * \partial net^{III}_{jk}}{\partial y^{II}_{1,j}} * \partial y^{II}_{1,j}}{\partial net^{III}_{1,j}}\right] = \sum_k \delta^{III}_{jk} y^{III}_{jk} \tag{4.11}$$

$$\delta^{II}_{2k} = \sum_j \left[\frac{\dfrac{\left(\dfrac{-\partial e_{idr}(t)\dot{e}_{idr}(t)}{\partial net^{III}_{jk}}\right) * \partial net^{III}_{jk}}{\partial y^{II}_{2,j}} * \partial y^{II}_{2,j}}{\partial net^{III}_{2,j}}\right] = \sum_k \delta^{III}_{jk} y^{III}_{jk} \tag{4.12}$$

The updated laws of: $m^{II}_{1,j}$, $m^{II}_{2,k}$, and $\sigma^{II}_{1,j}$, $\sigma^{II}_{2,k}$ also can be obtained by the gradient decent search algorithm:

$$\Delta m_{1,j}^{ll} = \frac{\frac{-\partial e_{idr}(t)\dot{e}_{idr}(t)}{\partial net_{1,j}^{ll}} * \partial net_{1,j}^{ll}}{\partial m_{1,j}^{ll}} = \mu_4 \frac{\delta_{1,j}^{ll} * 2 * \left(x_{1,j}^{ll} - m_{1,j}^{ll}\right)}{\left(\sigma_{1,j}^{ll}\right)^2} \qquad (4.13)$$

$$\Delta m_{2,k}^{ll} = \frac{\frac{-\partial e_{idr}(t)\dot{e}_{idr}(t)}{\partial net_{2,k}^{ll}} * \partial net_{2,k}^{ll}}{\partial m_{2,k}^{ll}} = \mu_3 \frac{\delta_{2,k}^{ll} * 2 * \left(x_{2,k}^{ll} - m_{2,k}^{ll}\right)}{\left(\sigma_{2,k}^{ll}\right)^2} \qquad (4.14)$$

$$\Delta \sigma_{1,j}^{ll} = \frac{\frac{-\partial e_{idr}(t)\dot{e}_{idr}(t)}{\partial net_{1,j}^{ll}} * \partial net_{1,j}^{ll}}{\partial \sigma_{1,j}^{ll}} = \mu_2 \frac{\delta_{1,j}^{ll} * 2 * \left(x_{1,j}^{ll} - m_{1,j}^{ll}\right)^2}{\left(\sigma_{2,k}^{ll}\right)^3} \qquad (4.15)$$

Where: μ_4, μ_3, μ_2 and μ_1 are the learning-rate parameters of the mean and the standard deviation of the Gaussian functions.

In this case, the *NFC* controller is proposed to control rotor *direct and quadrature* currents respectively (I_{rd} and I_{rq}) instead the *PID* controller, in order to improve the dynamic responses against parameter variation. The simulation results of the proposed algorithms are presented in Figs. **(4.16, 4.17)** and **4.19)**.

6. SIMULATION RESULTS

The proposed system (*DFIG control + wind turbine*)[3] is exactly the same used in chapter 3;

- The first part depicts the behavior of the wind-system parameters under three (03) modes –knowing that for each mode we present the different proposed topologies as follows:
- **1- Mode 1 (Red color/** Figs. **(4.4** and **4.16)**: Without MPPT Strategy, in this case we impose the P_s and Q_s reference profiles.
- **2- Mode 2 (Blue color/** Fig. **4.17)**: With MPPT strategy, in this case we propose a low wind speed based on step form (Max wind speed = 11.5 m/sec) by keeping stator reactive power equal to Zero level "Q_s = 0 (Var)"; to ensure only the exchange of the stator active power to the grid; means following the maximum active power point.
- **3- Mode 3 (Green color/** Fig. **4.19)**: With MPPT strategy, in this case we propose a medium wind speed based on random form (Max wind speed = 13.5 m/sec) by keeping stator reactive power equal to Zero level "Q_s = 0 Var"; to ensure only the exchange of the stator active power to the grid; means following the maximum active power point.
- The second part focuses on the comparative simulation study between the *MPPT* strategy for novel *IDPC* algorithms (*using T1-FLC, T2-FLC & NFC respectively*) using step' and random wind speed, described in detail in Fig.

(**4.18**) and **4.20**) respectively. This section is developed in order to illustrate the *Sub- and Super-synchronous* modes and the behavior of slip under generator speed variation.

- The third part deals with robustness tests using a comparative simulation study for three modes (*with/without MPPT strategy*) described in detail in Figs. **4.21**, **4.22** and **4.23**) respectively. This section is developed in order to verify the robustness of wind-system under parameter variation (*using three tests*)[4] in transient *and* steady states.

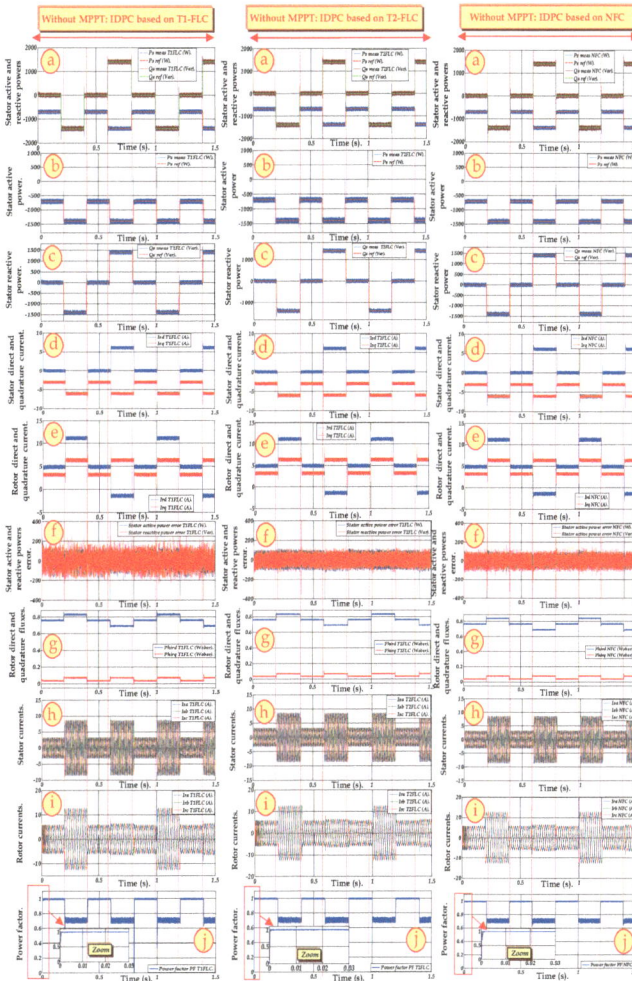

Fig.(4.16). Simulation results of **Mode-1** using T1-FLC, T2-FLC & NFC; (**a**): stator active and reactive powers, (**b**): stator active power, (**c**): stator reactive power, (**d**): stator direct and quadrature currents, (**e**): rotor direct and quadrature currents, (**f**): stator active and reactive power error, (**g**): rotor direct and quadrature fluxes, (**h**): stator currents, (**i**): rotor currents, (**j**): power factor.

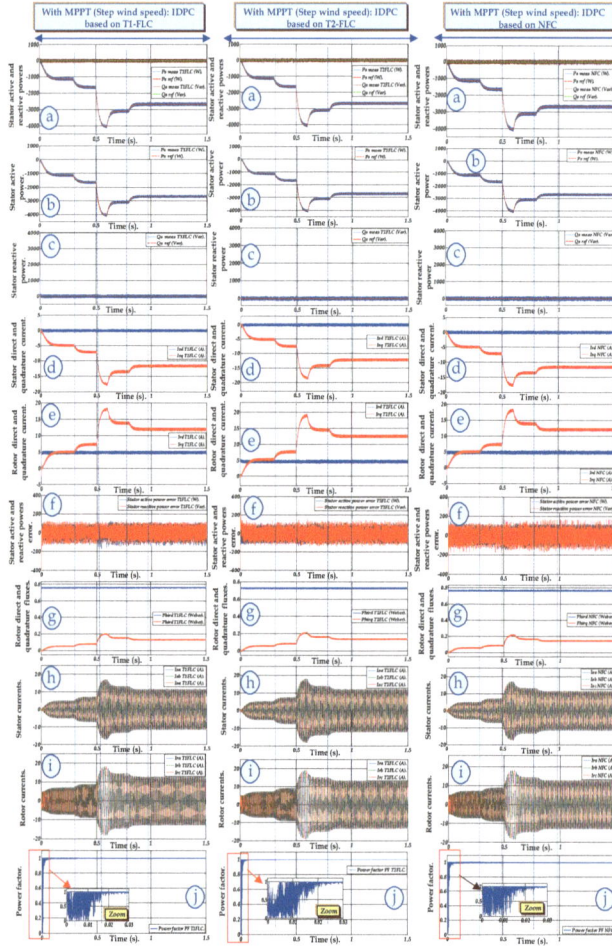

Fig. (4.17). Simulation results of **Mode-2** using T1-FLC, T2-FLC & NFC; (**a**): stator active and reactive powers, (**b**): stator active power, (**c**): stator reactive power, (**d**): stator direct and quadrature currents, (**e**): rotor direct and quadrature currents, (**f**): stator active and reactive power error, (**g**): rotor direct and quadrature fluxes, (**h**): stator currents,

Table. 4.4 The proposed profiles of the active and reactive power references.

Time (sec):	Stator active power (W):	Stator reactive power (Var):
[0 - 0.2]	-700.	0.
[0.2–0.4]	-1400.	-1400.
[0.4–0.6]	-700.	0.
[0.6–0.8]	-1400.	+1400.
[0.8–1.0]	-700.	0.
[1.0–1.2]	-1400.	-1400.
[1.2–1.4]	-700.	0.
[1.4–1.5]	-1400.	+1400.

6.1. Mode 1 (Based on T1-FLC, T2-FLC 3x0026; NFC, Without MPPT Strategy)

A-Novel IDPC based on T1-FLC: (Fig. 4.16 to the left side):

It is clear that the measured powers (active and reactive) have bad tracking power (big error) compared to their reference powers in transient and steady states (refer to Fig. **4.16**-(**b** and **c**) to the left side), a few ($\approx$ 10%) overshoot is noted at 0.2 (sec) and 0.8 (sec) respectively in measured stator active' and reactive power (P_{s_meas} and Q_{s_meas}). We observe a small power error of active and reactive powers -100 (W_Var) $\leq \Delta P_s\_\Delta Q_s \leq$ +100 (W_Var). Table (**4.4**) depicts in details the obtained results of novel IDPC based on T1-FLC.

B-Novel IDPC based on T2-FLC: (Fig. 4.16 to the middle side):

The measured powers (P_{s_meas} and Q_{s_meas}) have good tracking power (refer to Fig. **4.16**-(**b** and **c**) *to the middle side*), a few ($\approx$ 9%) overshoot is noted at 0.2 (sec) and 0.8 (sec) respectively in P_{s_meas} and Q_{s_meas}-*because the T2-FLC controller is robust in terms of overshoot especially if the step power' changement is big-*. We observe few power errors: -100 (W_Var) $\leq \Delta Ps\_\Delta Qs \leq$ +100 (W_Var), a good THD of stator currents will be injected into the grid *(= 01.07% respects the IEEE-519 Std)*. The obtained results of novel IDPC based on T2-FLC; are described in details in Table **4.4**.

C-Novel IDPC based on NFC: (Fig. 4.16 to the right side):

It is noted that the measured powers have good tracking power (refer to Fig. **4.16**-(**b** and **c**) to the right side), a few (9%) overshoot is noted is noted in interval: 0.2 and 0.8 (sec) respectively in P_{s_meas} and Q_{s_meas}-*because the NFC controller is robust in terms of overshoot -*. A good THD of stator currents will be injected into the grid *(= 01.19% respect the IEEE-519 Std)* and a small power errors are noted: -120 (W_Var) $\leq \Delta P_s\_\Delta Q_s \leq$ +120 (W_Var). The obtained results of novel IDPC based on NFC; are shown in details in Table (**4.4**).

Table. 4.5 Performances results for proposed control based on Mode 1 using T1-FLC, T2-FLC NFC.

	THD_Is_abc (%):	THD_Ir_abc (%):	Overshoot :	Response time (sec) :	Power Error (W_Var) :
T1-FLC :	01.20 %	41.45 %	Few ($\approx$ 10%).	$0.8 * 10^{-3}$.	+/- 100.
T2-FLC :	01.07 %	40.48 %	Few ($\approx$ 9%).	$0.8 * 10^{-3}$.	+/- 100.
NFC :	01.19 %	88.20 %	Few ($\approx$ 9%).	$0.33 * 10^{-3}$.	+/- 120.

6.2. Mode 2 (Based on T1-FLC, T2-FLC NFC, with MPPT Strategy- Step Wind Speed)

A- *Novel IDPC based on T1-FLC:* (Fig. 4.17 to the left side):

It is noted that the poor power tracking of the measured powers compared to their reference powers in transient and steady states (refer to Fig. **4.17**-(**b** and **c**)), a neglected ($\approx$ 7%) overshoot is noted at 0.6 (sec) in measured stator active' and reactive power (P_{s_meas} and Q_{s_meas}). We observe a small power error of active and reactive powers -130 (W_Var) $\leq \Delta P_s\_\Delta Q_s \leq$ +130 (W_Var). Table. 4.6 shows in details the obtained results of the novel IDPC based on T1-FLC.

B- *Novel IDPC based on T2-FLC:* (Fig. 4.17 to the middle side):

It is clear that the measured powers (active and reactive) have good tracking power compared to their reference powers in transient and steady states (refer to Fig. **4.17**-(**b** and **c**)) to the left side), a neglected (6%) overshoot is noted at 0.6 (sec) in P_{s_meas} and Q_{s_meas}-*because T2-FLC controller is robust in terms of overshoot especially if the step power' changement is big-*. We observe a low power error of active and reactive powers -100 (W_Var) $\leq \Delta P_s\_\Delta Q_s \leq$ +100 (W_Var). Table **4.6** depicts in details the obtained results of the novel IDPC based on T2-FLC.

C- *Novel IDPC based on NFC:* (Fig. 4.17 to the right side):

The measured powers (active and reactive) have good tracking power compared to their reference powers in transient and steady states, a neglected (5%) overshoot is noted at 0.6 (sec) in P_{s_meas} and Q_{s_meas}-*because the NFC controller is very robust in terms of overshoot especially if the wind power' changement is big-*. We observe an acceptable power error of active and reactive powers -150 (W_Var) $\leq \Delta Ps\_\Delta Qs \leq$ +150 (W_Var). We remark the sinusoidal form of the waveforms of the stator' and rotor currents; I_{s_abc} and I_{r_abc} and excellent THD of stator currents will be injected into the grid *(= 0.60% respect the IEEE-519 Std)*. Table **4.6** described in details the obtained results of the novel IDPC based on NFC.

MPPT Strategy[6]: Knowing that in this case (**Mode: 2**) the wind speed had taken the step form.

A- *Novel IDPC based on T1-FLC:* (Fig. 4.18 to the left side):

A good power tracking of the measured powers is noted compared to their reference powers in transient and steady states (refer to Fig. **4.18**a), it can be seen few ripples and undulation especially in steady states for the P s_meas and Q s_meas (Fig. **4.18a-c**).

Table. 4.6 Performances results for proposed control based on Mode 2 using T1-FLC, T2-FLC NFC.

	THD_Is_abc (%):	THD_Ir_abc (%):	Overshoot :	Response time (sec) :	Power Error (W_Var) :
T1-FLC :	0.67 %	05.88 %	Neglected ($\approx$ 7%).	$0.7 * 10^{-3}$.	+/- 130.
T2-FLC :	0.68 %	05.99 %	Neglected (6%).	$0.67 * 10^{-3}$.	+/- 100.
NFC :	0.60 %	111.70 %	Neglected (5%).	$0.37 * 10^{-3}$.	+/- 150.

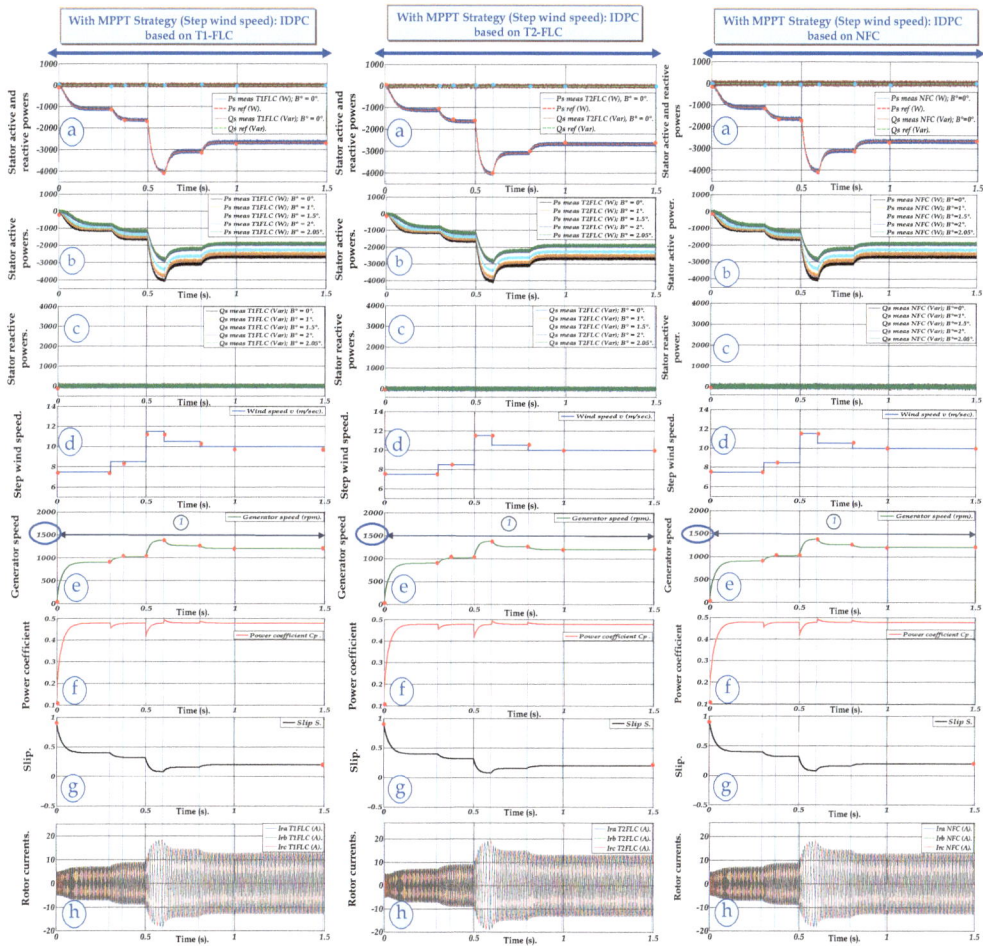

Fig. (4.18). Simulations results of two MPPT strategies based on Mode 2 (using T1-FLC, T2-FLC & NFC respectively); (**a**): stator active and reactive powers, (**b**): stator active power using different B° pitch angles, (**c**): stator reactive power using different pitch angles B°, (**d**): wind speed, (**e**): generator speed, (**f**): power coefficient, (**g**): slip, (**h**): rotor currents.

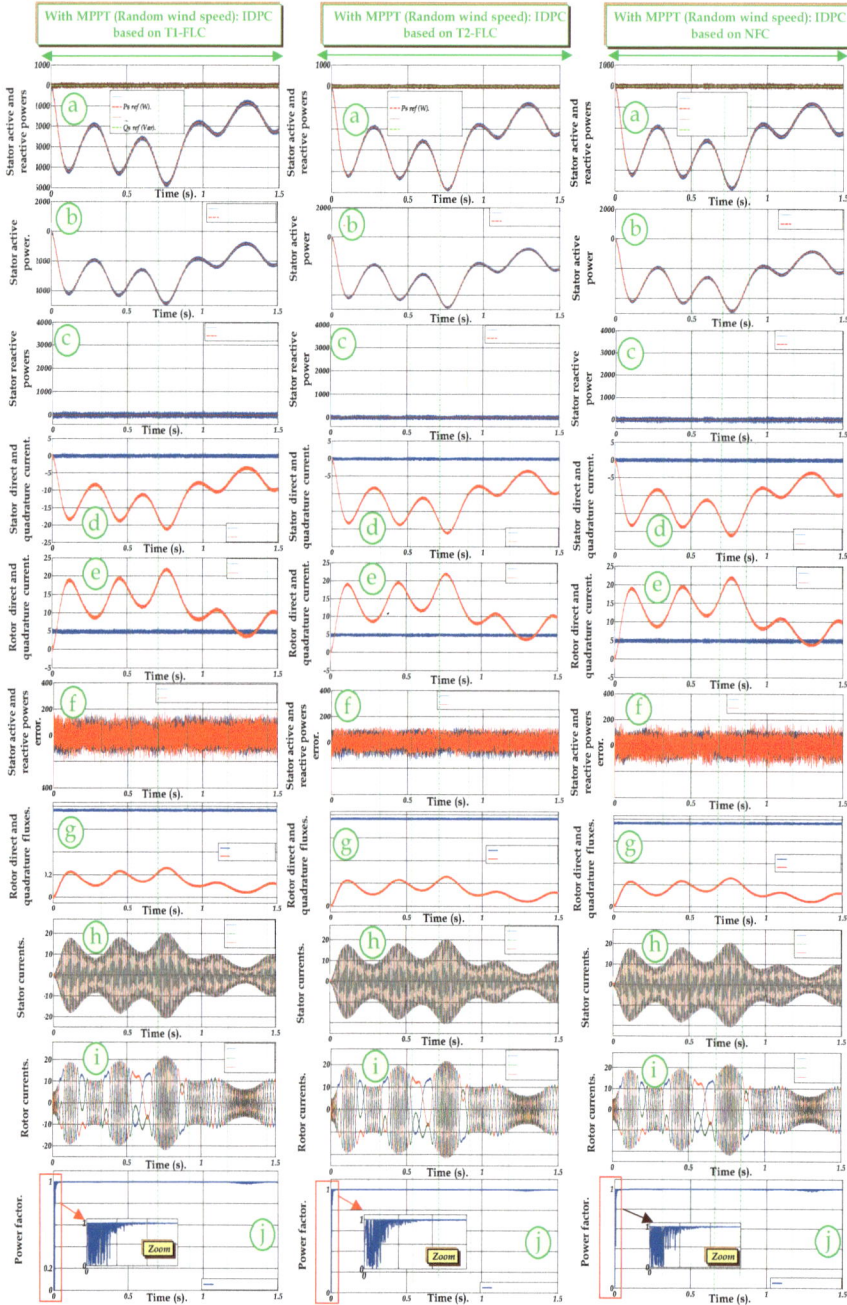

Fig.(4.19). Simulation results of **Mode-3** using T1-FLC, T2-FLC NFC; (**a**): stator active and reactive powers, (**b**): stator active power, (**c**): stator reactive power, (**d**): stator direct and quadrature currents, (**e**): rotor direct and quadrature currents, (**f**): stator active and reactive power error, (**g**): rotor direct and quadrature fluxes, (**h**): stator currents, (**i**): rotor currents, (**j**): power factor.

B-Novel IDPC based on T2-FLC: (Fig. 4.18 to the middle side):

An excellent power tracking of the measured powers is noted compared to their reference powers in transient and steady states (refer to Fig. (**4.18-(a, b c)**), prove that T2-FLC controller offer good tracking power especially if in wind speed severe variation compared to T1-FLC.

C-Novel IDPC based on NFC: (Fig. 4.18 to the right side):

It is clear that the measured powers follow exactly their reference powers in transient and steady states (refer to Fig. (**4.18-(a, b c)**) which present an excellent power tracking, and prove also that the proposed NFC controller offer an excellent wind-system performance in severe conditions compared to the conventional fuzzy controller (Type-1).

6.3. Mode 3 (Based on T1-FLC, T2-FLC NFC, with MPPT Strategy- Random Wind Speed)

A-Novel IDPC based on T1-FLC: (Fig. 4.19 to the Left Side):

It is clear that the measured powers (active and reactive) have good tracking, a few ($\approx 10\%$) overshoot is noted at 0.5 (sec) and 0.8 (sec) in measured stator reactive' and active power (P_{s_meas} and Q_{s_meas}/refer to 4.19-(b and c). We observe an acceptable power error of active and reactive powers: -150 (W_Var) $\leq$ $\Delta P_s\_\Delta Q_s \leq +150$ (W_Var). Table (**4.7**) describes in details the obtained results of the improved IDPC based on T1-FLC.

B-Novel IDPC based on T2-FLC: (Fig. 4.19 to the Middle Side):

A good tracking power is noted between reference and measured powers in transient and steady states (refer to Fig. (**4.19-(b and c)**), a neglected (6%) overshoot is noted at 0.5 (sec) and 0.8 (sec) in measured stator reactive' and active power (P_{s_meas} and Q_{s_meas}) (*because the T2-FLC controller is robust in terms of overshoot especially if the wind power' changement is big*). We observe a lower power error of active and reactive powers -120 (W_Var) $\leq \Delta P_s\_\Delta Q_s \leq +120$ (W_Var). We remark the sinusoidal form of the waveforms and excellent THD of stator currents which will be injected into the grid (*= 0.60% for stator currents and 04.90% for rotor currents respectively respect the IEEE-519 Std*). Table (**4.7**) depicts in details the obtained results of the proposed IDPC based on T2-FLC.

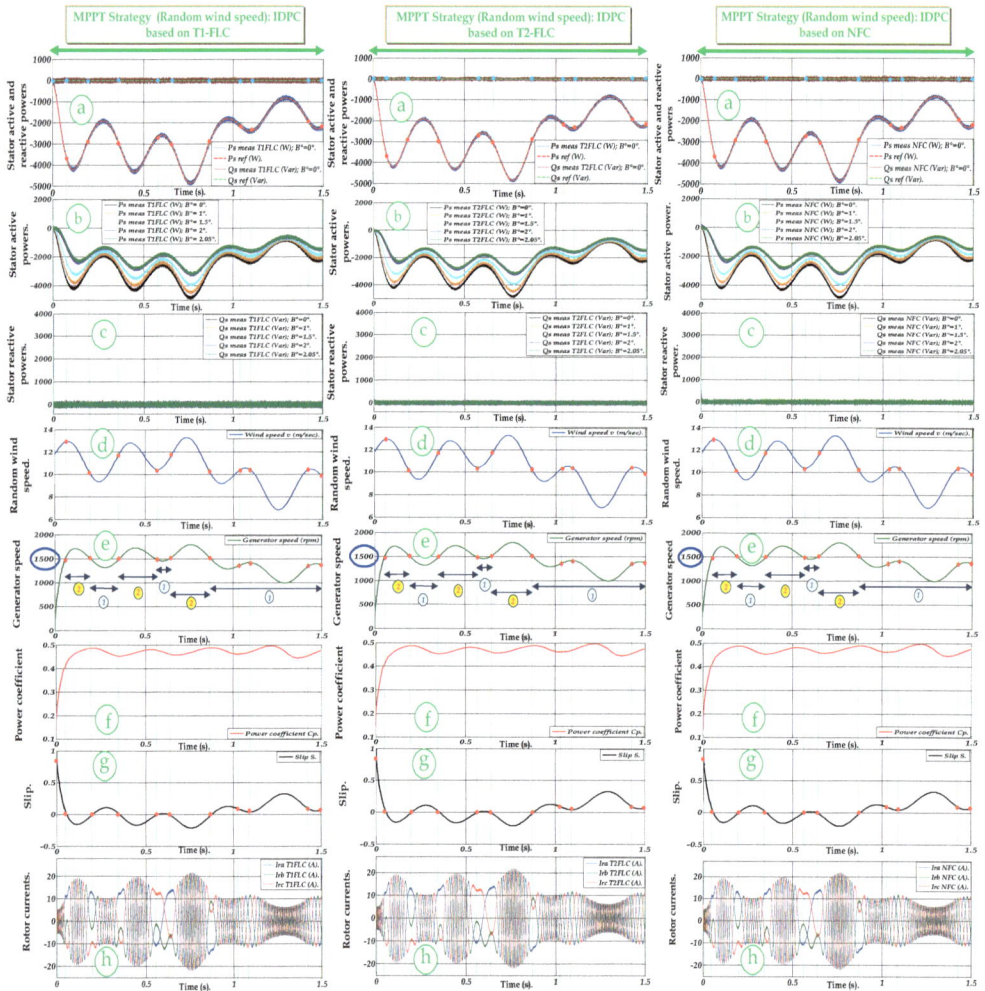

Fig.(4.20). Simulations results of two MPPT strategies based on **Mode 3** (using T1-FLC, T2-FLC NFC respectively); **(a)**: stator active and reactive powers, **(b)**: stator active power using different B° pitch angles, **(c)**: stator reactive power using different pitch angles B°, **(d)**: wind speed, **(e)**: generator speed, **(f)**: power coefficient, **(g)**: slip, **(h):** rotor currents.

Table. 4.7 Performances results for proposed control based on Mode 3 using T1-FLC, T2-FLC NFC.

	THD_Is_abc (%):	THD_Ir_abc (%):	Overshoot :	Response time (sec) :	Power Error (W_Var) :
T1-FLC :	0.44%	06.54 %	Few (≈ 10%).	$0.42 * 10^{-3}$.	+/- 150.
T2-FLC :	0.60 %	04.90 %	Neglected (6%).	$0.75 * 10^{-3}$.	+/- 120.
NFC :	0.56 %	05.26 %	Neglected (5%).	$0.37 * 10^{-3}$.	+/- 150.

C-Novel IDPC based on NFC: (Fig. 4.19 to the Right Side):

An excellent tracking power is noted in transient and steady states, a neglected (5%) overshoot is noted at 0.5 (sec) and 0.8 (sec) in measured stator reactive' and active power (P_{s_meas} and Q_{s_meas}), -*because the NFC controller is robust in terms of overshoot especially if the step power' changement is big-*. An excellent THD of stator currents will be injected into the grid (= *0.56% for stator currents and 05.26% for rotor currents respectively respect the IEEE-519 Std*). An acceptable power error is noted: -150 (W_Var) $\leq \Delta P_s\_\Delta Q_s \leq$ +150 (W_Var). Table **4.7** shows in details the obtained results of the novel IDPC based on NFC.

MPPT Strategy[7]: in this case (*Mode: 3*); the wind speed had taken the random form.

A-Novel IDPC based on T1-FLC: (Fig. 4.20 to the Left Side):

We note good power tracking of the measured powers compared to their reference powers in transient and steady states (refer to Fig. **4.20-(a))**, as few ripples and undulation can be seen especially in steady states for the P_{s_meas} and Q_{s_meas} (Fig. **4.17-(a, b c))**.

B-Novel IDPC based on T2-FLC:> (Fig. 4.20 to the Middle Side):

An excellent power tracking of the measured powers is noted compared to their reference powers in transient and steady states (refer to Fig. **4.20-(a, b c))**. Neglected ripples can also be seen which prove that T2-FLC controller offer good tracking power especially if in wind speed severe variation compared to T1-FLC.

C-Novel IDPC based on NFC: (Fig. 4.20 to the Right Side):

It is noted that the measured powers follow exactly their reference powers in transient and steady states (refer to Fig. **4.20-(a, b c))**, an excellent wind-system performances are noted in term: response time (0.37 * 10.e^{-3} (sec)) and overshoot (5%).

6.4. Robustness Tests[7] for Mode 1, Mode 2 Mode 3

Fig. **4.**(**21**.(a,b)-**22**.(a,b)-**23**.(a,b)) illustrate the behavior of measured active and reactive powers and theirs references respectively under parameters variations; in transient and steady states.

A-Mode 1 (Novel IDPC based on T1-FLC, T2-FLC NFC):

It can be noted few power in active and reactive power (topology based on T1-FLC/please refer to the left side of Fig. **4.21**) especially using the 2nd test and 3rd test (green color) with small undulations especially in transient and steady states (*please refer to zoom*) and the value of power error reaches nearly $\pm$ 150 (W_Var) in Test-1, and nearly $\pm$220 (W_Var) in Test-2 Test-3.

For the second topology (based on T2-FLC/refer to the middle side of Fig. **4.21**) a lower power error is noted for the Test-1 (blue color) reaches nearly $\pm$ 120 (W_Var), and nearly $\pm$ 180 (W_Var) in Test-2 and Test-3.

For the third topology (based on NFC/ refer to the right side of Fig. **4.21**) also a small power error is oted fort the first Test (blue color), and the value of power error reaches nearly $\pm$ 150 (W_Var) in Test-1, and nearly $\pm$175 (W_Var) in Test-2 and Test-3. A few overshoot is noted under all robustness tests especially at 0.2 (sec) and at 1.0 (sec).

B-Mode 2 (Novel IDPC based on T1-FLC, T2-FLC NFC):

Fig. (**4.22**)-(based on T1-FLC/please refer to the left side) display the behavior of stator active and reactive powers under MPPT strategy by maintaining the reactive power equals to zero value. In this case the active power had taken the inverse step profile of wind speed. Using robustness tests a big undulations are noted (using tests: 2 and 3) especially at 0.6 (sec) which presents the rated power of DFIG (P=4 (kW)), on other hand and in the same time a remarkable power error is noted in stator reactive power, also a big overshoot is noted in steady state of active and reactive power.

For the second topology (based on T2-FLC/ refer to middle side of the Fig. **4.22**) the active power had taken the inverse step profile of wind speed. Using robustness tests a few undulations are noted, on other hand and in the same time a limited power error is noted in stator reactive.

In the third topology (based on NFC/ refer to right side of the Fig. **4.22**) a neglectd ripples are noted (using tests: 2 and 3) especially at 0.6 (sec) which presents the rated power of DFIG, and in the same time a very small power error is noted in stator reactive power (*which means that the NFC can maintain the unity power factor under parameters variation*), also a lower overshoot is noted in transient and steady states of P_{s_meas} $Q_{s\_meas.}$

C-Mode 3 (Novel IDPC based on T1-FLC, T2-FLC NFC):

Fig. (**4.23**)-(topology based on T1-FLC/refer to the left side) display the behavior of stator active and reactive powers under MPPT strategy by maintaining the reactive power equals to zero value. In this case the active power had taken the inverse random profile of wind speed. Using robustness tests a big undulations are noted (using tests: 2 and 3) especially at 0.75 (sec) and 0.8 (sec) which presents the over rated power of DFIG (P=4 (kW) and the measured active power maintain 4.6 (kW)), also a high overshoot is noted in transient and steady states.

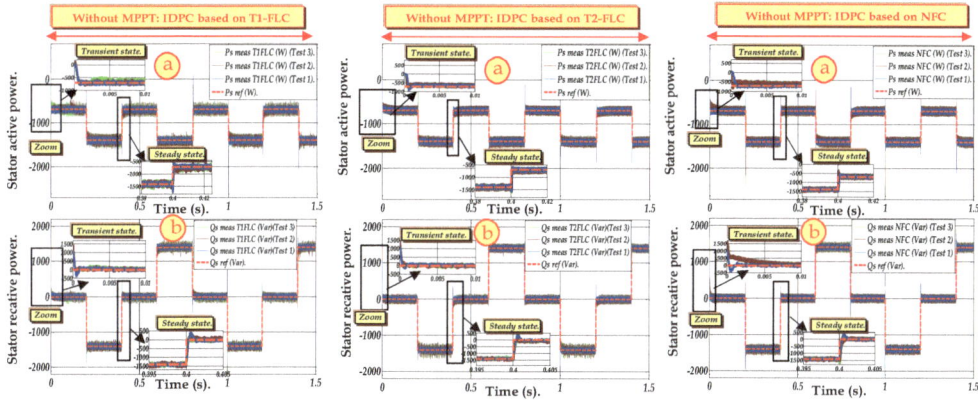

Fig.(4.21). Robustness tests of **Mode 1** for proposed control using T1-FLC, T2-FLC NFC respectively; (**a**): stator active powers, (**b**): stator reactive powers.

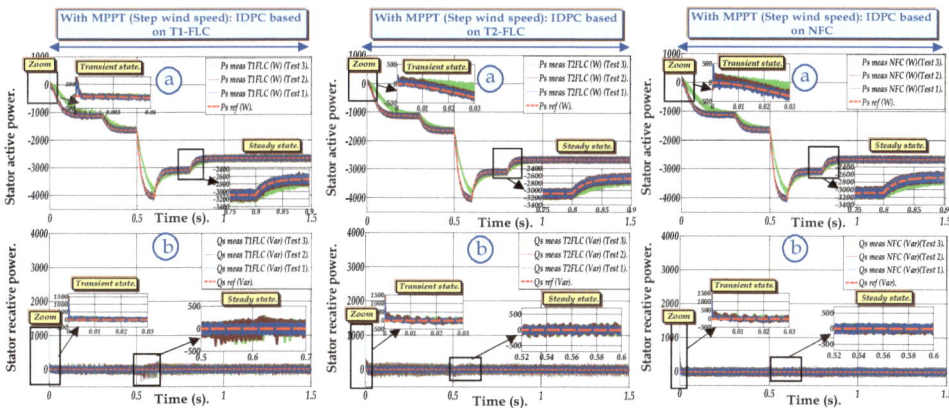

Fig.(4.22). Robustness tests of **Mode 2** for proposed control using T1-FLC, T2-FLC NFC respectively; (**a**): stator active powers, (**b**): stator reactive powers.

In the second topology (based on T2-FLC/refer to the middle side of the Fig. **4.23**) a few undulations are noted (using tests: 2 3) which presents the over rated power of DFIG, in the same time a few power error is noted in stator reactive power, also a few overshoot is noted in transient and steady states of active and reactive power (*refer to the zoom*).

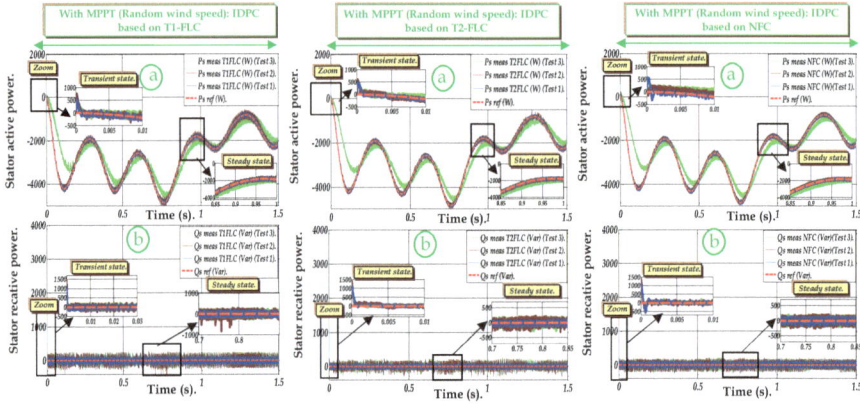

Fig.(4.23). Robustness tests of **Mode 3** for proposed control using T1-FLC, T2-FLC NFC respectively; (**a**): stator active powers, (**b**): stator reactive powers.

For the third topology (based on NFC/refer to right side of the Fig. **4.23**) the neglected ripples are noted (using tests: 2 3), an acceptable power error and overshoot (*which means that the NFC controllers can maintain the unity power factor under parameters changing*) are shown in stator reactive power especially in steady states (*Please refer to the zoom*).

7. WIND-SYSTEM PERFORMANCES RECAPITULATION UNDER SIX (*06*) PROPOSED *IDPC* ALGORITHMS

Table 4.8. Wind-system performances (under six proposed IDPC algorithms).

		Overshoot:	Response time:	THD:	Power error:	Power tracking:	Sensitivity of the parameters changement:
Chapter 3:	Topology 1 (*IDPC based on PI*):	--	+	--	--	-	--
	Topology 2 (*IDPC based on PID*):	+	+	+	-	+	-
	Topology 3 (*IDPC based on MRAC*):	++	+	+	+	++	+
Chapter 4:	Topology 4 (*IDPC based on T1-FLC*):	+	++	++	++	++	++
	Topology 5 (*IDPC based on T2-FLC*):	++	++	+++	+++	+++	++
	Topology 6 (*IDPC based on NFC*):	++	+++	+++	+++	+++	+++

NB: "-" means Poor performance and "+" means High performance.

In this section, we present the advantages and disadvantages of each proposed algorithms (*Chapter: 03 and Chapter: 04*) using six (*06*) performance criteria, taken into account three (*03*) proposed modes *and* robustness tests.

After analyzing the recapitulation results (as demonstrates in Table **4.8**), it is clear that the artifical intelligent controllers offer an improved' wind-system performances in transient and steady states, and we can confirm also that the best artifical intelligent controllers are T2-FLC and NFC.

CONCLUSION

In this chapter, improved *IDPC* algorithms for *DFIG*-grid connection have been proposed. In order to control independently *DFIG's* stator powers; novel *IDPC* with *SVM* have been combined to adjust active and reactive powers and rotor currents. *MPPT* strategy was proposed in order to extract the maximum wind-power despite the sudden wind speed variation (*in this chapter we proposed two wind speed profiles: the step wind speed and the random wind speed*) and to maintain the unity power factor (*PF ≈ 1*). In order to enhance the conventional *IDPC*; *T1-FLC, T2-FLC and NFC* were proposed instead the *PID* controllers to control I*rd* and I*rq* respectively. In this context, several drawbacks in transient and steady states were treated using the adaptive and intelligent controllers, in terms of tracking power and dynamic response performances. The simulation results have been developed *via Matlab/ Simulink®* software, illustrate high dynamics response and improved wind-system performances regardless wind-speed variation. Using the robustness tests, the wind-performances are remarkably improved (especially for Topology: 5 and Topology: 6 from Table **4.9**) which demonstrate the high performance ability of artificial intelligent algorithms (*T2-FLC and NFC*) to keep the power tracking trajectory under parameter changing and sudden wind speed variation. In the next chapter, a new non-linear control will be proposed in order to overcome the coupling terms (*under d-q axes*).

NOTES

[1] Knowing that; the same linguistic rules (for T1-FLC) will be applied for the rotor quadrature current Irq.

[2] Knowing that; the same linguistic rules (for T2-FLC) will be applied for the rotor quadrature current Irq.

[3] Please refer to Appendix A :(The DFIG' and wind turbine parameters are indicated in Table **A.1** and Table **A.2** respectively).

[4] Knowing that in this chapter the robustness tests are based on three tests as follows: [Test-1: without parameter changement → Blue color, Test-2: +100% of Rr and -25% of (L_s, L_r and L_m) → Brown color and Test-3: +100% of (J and R_r), -25 % of (L_r, L_s and L_m) → Green color] respectively.

[5] Please refer to section §.3.10.2 for more information.

[6] Please refer to section §.3.11.2 for more information.

[7] Knowing that in this chapter the robustness tests are based on three tests as follows: [Test-1: without parameter changement → Blue color, Test-2: +100% of Rr and -25% of (Ls, Lr and Lm) → Brown color and Test-3: +100% of (J and Rr), -25 % of (L_r, L_s and L_m) → Green color] respectively.

REFERENCES

[1]   M.B. Ozek, and Z.H. Akpolat, *A Software tool: Type-2 fuzzy logic toolbox.* vol. Vol. 22. JWPI, 2008.

[2]   S. Mikkili, and A.K. Panda, *Review and Analysis of Type-1 and Type-2 Fuzzy Logic Controllers.* vol. Vol. 08. JASE, 2014.

[3]   T. Ramesh, A. Kumar Panda, and S. Shiva Kumar, "Type-2 fuzzy logic control based MRAS speed estimator for speed sensorless direct torque and flux control of an induction motor drive", *ISA Trans.,* vol. 57, pp. 262-275, 2015.
       [http://dx.doi.org/10.1016/j.isatra.2015.03.017] [PMID: 25887841]

[4]   N.V. Naik, and S.P. Singh, "A Novel Type-2 Fuzzy Logic Control of Induction Motor Drive using Scalar Control", *IEEE Conference, 5th India International Conference on Power Electronics (IICPE),* 2012

[5]   Q. Liang, and J.M. Mendel, *Interval Type-2 Logic Systems: Theory and Design,* 2000.

[6]   A. Chaiba, R. Abdessemed, and M.L. Bendaas, "Hybrid Intelligent Control based Torque Tracking approach for Doubly Fed Asynchronous Motor (DFAM) drive", *Journal of Electrical Systems,* vol. 8, no. 3, pp. 262-272, 2012.

[7]   C. Elmas, O. Ustun, and H.H. Sayan, "A neuro-fuzzy controller for speed control of a permanent magnet synchronous motor drive", *Expert Syst. Appl.,* vol. 34, no. 1, pp. 657-664, 2008.
       [http://dx.doi.org/10.1016/j.eswa.2006.10.002]

[8]   M. Gökbulut, B. Dandil, and C. Bal, *A hybrid neuro-fuzzy controller for brushless DC motors.* Artificial Intelligence and Neural Networks, 2006, pp. 125-132.
       [http://dx.doi.org/10.1007/11803089_15]

[9]   K. Kouzi, M. Nait-Said, M. Hilairet, and É. Berthelot, *A fuzzy sliding-mode adaptive speed observer for vector control of an induction motor.* IEEE Int Conf. Indust. Elect IECON, 2008.

[10]   R. Marchi, P. Dainez, F.V. Zuben, and E. Bim, "A multilayer perceptron controller applied to the direct power control of a doubly fed induction generator", *IEEE Trans. Sustain. Energ.,* vol. 5, no. 2,

pp. 498-506, 2014.
[http://dx.doi.org/10.1109/TSTE.2013.2293621]

[11]   M.G. Simoes, B.K. Bose, and R.J. Spiegel, ""Design and Performance Evaluation of a Fuzzy-Logi-
        -Based Variable-Speed Wind Generation System", Chapter-1, pp.1-35, Title Book: "Fuzzy Control
        Systems, Analysis and Performances Evaluation", *IEEE Trans. Ind. Appl.*, vol. 33, no. 4, 1997.
        [http://dx.doi.org/10.1109/28.605737]

[12]   Fuzzy Logic With Engineering Applications.*Book.* 2nd ed. John Wiley Sons Ltd: England, 2004.

[13]   T.J. Ross, Fuzzy Logic With Engineering Applications.*Book.* 3rd ed. John Wiley Sons Ltd: United
        Kingdom, 2010.

[14]   K.M. Passino, and S. Yurkovich, *"Fuzzy Control", Book.* 1st ed. Addison-Wesley Longman, Inc:
        California, USA, 1998.

[15]   F. Amrane, and A. Chaiba, "Direct Power Control for grid-connected DFIG using Fuzzy Logic with a
        Fixed Switching Frequency", *Conférence Internationale d'Automatique et de la Mécatronique, CIAM
        10-11th Nov 2015 Oran-Algeria.*

[16]   G. Chen, and T.T. Pham, Introduction to Fuzzy Sets, Fuzzy Logic, And Fuzzy Control Systems.*Book.*
        1st ed. CRC Press: New York, USA, 2001.

[17]   A. Kouadria, T. Allaoui, M. Denaï, and G. Pissanidis, "Grid Power Quality Enhancement Using Fuzzy
        Control-Based Shunt Active Filtering", *SAI Intelligent Systems Conference,* November 10-11, 2015
        London, UK.
        [http://dx.doi.org/10.1109/IntelliSys.2015.7361208]

[18]   D. WU and J. M. Mendel, "Designing Practical Interval Type-2 Fuzzy Logic Systems Made Simple",
        2014 IEEE International Conference on Fuzzy Systems (FUZZ-IEEE)., 2014 Beijing, China.

[19]   F-J. Lin, *Introduction to Type-2 Fuzzy Logic Control: Theory and Applications.* Book, IEEE Systems,
        Man, Cybernetics Magazine, 2015.

[20]   S.K. Raju, and G.N. Pillai, *Design and Implementation of Type-2 Fuzzy Logic Controller for DFIG-
        Based Wind Energy Systems in Distribution Networks,* 2016.

[21]   L. Suganthia, S. Iniyan, and A.A. Samuel, *Applications of fuzzy logic in renewable energy systems – A
        review.* vol. 48. JRSER, 2015.

[22]   F. Amrane, and A. Chaiba, "Type-2 Fuzzy Logic Control: Design and Application in Wind Energy
        Conversion System based on DFIG *via* Active and Reactive Power Control", In: *Fuzzy Control
        Systems, Analysis and Performances Evaluation* Nova Science Publishers: USA, 2017, pp. 1-35.

[23]   F. Amrane, A. Chaiba, and B. Francois, "Application of Adaptive T2FLC in Stator Active and
        Reactive power Control WECS based on DFIG *via* Hypo/Hyper-Synchronous Modes", *4ième
        Conférence des Jeunes Chercheurs en Génie Electrique, JCGE, 30 Mai et 1er Juin 2017, Arras,
        France..*

[24]   F. Amrane, A. Chaiba, and B. Francois, "Suitable Power Control based on Type-2 Fuzzy Logic
        Control for Wind-Turbine DFIG Under Hypo-Synchronous Mode Fed by NPC Converter", *5th
        International Conference on Electrical Engineering, IEEE Conference, ICEE 29-31th Oct 2017
        Boumerdes-Algeria.*

[25]   F. Amrane, and A. Chaiba, "A Hybrid Intelligent Control based on DPC for grid-connected DFIG with
        a Fixed Switching Frequency using MPPT Strategy", *4th International Conference on Electrical
        Engineering, IEEE Conference, ICEE 13-15th Dec 2015 Boumerdes-Algeria.*
        [http://dx.doi.org/10.1109/INTEE.2015.7416678]

CHAPTER 5

General Conclusion

This Book deals with the enhancement of the power control of WECS based on DFIG. The conventional IDPC is a control method which offers a decoupled active and reactive power control for wind system drives. IDPC is featured by simple structure, fast power dynamic. However, it suffers from the high sensitivity to the machine parameters variation and wind speed variation.

Due to the non-linearity of wind system, the power control of DFIG presents a big challenge under wind-speed variation and sensibility parameter. To overcome these problems an improved IDPC (based on classical PID controller) was proposed, in order to enhance the wind-system performances in terms of power error, tracking power and overshoot. Unfortunately using robustness tests (*via* severe tests); the wind-system offers non-satisfactory simulation results which were illustrated by the very bad power tracking and very big overshoot (>50%). In this context; adaptive, robust and intelligent controllers were proposed to control direct and quadrature currents (I_{rd} and I_{rq}) under MPPT strategy. In this case, the new IDPC based on intelligent controllers offered an excellent wind-system performance especially using robustness tests, which offered a big improvement especially using T2-FLC and NFC.

The main objective of this Book is the improvement of the WECS' performances based on DFIG' drive, under power control using various control approaches. In this context, the research work has addressed these principal points concerning the conventional control algorithm' drawbacks:

1. The parameters changement sensibility (noted in conventional IDPC) is solved using intelligent and robust controllers (Improved IDPC).
2. The important overshoot noted especially in conventional IDPC is overcome using adaptive and artificial controllers (especially using T2-FLC and NFC).
3. The response time is improved (from the order of $10e^{-2}$ (sec) to $10e^{-3}$ (sec)).
4. The power tracking also is more improved especially using the improved IDPC.

5. Good power/voltage quality which will be transmitted to the grid (excellent THD that respect IEEE standards (< + 5%)) and Power factor is maintained to the unitary (PF≈1) despite the wind-system variation.
6. A neglected power error less than 100 (W_Var) presents < 2.5 % of the rated power.

5.1. FUTURE WORKS

- The experimental implementation ground by the use of the dSPACE 1104 signal card.
- The load variation is interesting study in the case of both proposed algorithms (FOC-HCC and IDPC) in order to ensure the power transmitted limit. Secondly, stator voltage control is necessary (at the load level) to improve the energy management at all times.
- Artificial intelligence (Fuzzy Control and Neuro-Fuzzy Control) is an adequate solution in an experimental implementation (using the same experimental test bench) to improve the stator powers and the rotor currents control, despite the sudden wind variation, parameters variation and load variation, respectively.
- Experimental implementation of the MPPT strategy (*via* Emulator turbine), in order to extract the maximum wind power despite the wind-speed variation by means of artificial intelligent controllers.
- In order to raise the output power of WECS, the 3L-NPC converters are the best solutions in experimental validation especially if the aerodynamic power is high.
- In order to ensure the accuracy of PI', PID' and MRAC' gains; Genetic algorithms (GA) and Particle Swarm Optimization (PSO) will be proposed to minimize the calculation error and to offer optimal values for the proposed algorithm.
- The predictive control is an excellent alternative for the Wind-system based on the independent control of the active and reactive powers. The basic principle of this command does not require the PWM modulation (contrary case of IDPC) which minimizes the calculation time and also it is easy to implement.
- In order to minimize the cost, the observers (speed …*etc*) will be interesting idea in experimental implementation instead the sensors.

APPENDIX A: WECS PARAMETERS

Table. A.1 Parameters of the DFIG.

Rated Power:	4.0 kW
Stator Resistance:	$Rs = 1.2\ \Omega$
Rotor Resistance:	$Rr = 1.8\ \Omega$
Stator Inductance:	$Ls = 0.1554\ H.$
Rotor Inductance:	$Lr = 0.1558\ H.$
Mutual Inductance:	$Lm = 0.15\ H.$
Rated Voltage:	$Vs = 220/380\ V$
Number of Pole pairs:	$P= 2$
Rated Speed:	$N=1440\ rpm$
Friction Coefficient:	$f_{DFIG} =0.00\ N.m/sec$
The moment of inertia	$J=0.2\ kg.m^{2}$

Table. A.2 Parameters of the wind turbine.

Rated Power:	4.5 kW
Number of blades:	$P= 3$
Blade diameter	$R= 3\ m$
Gain:	$G=4.15$
The moment of inertia	$J_{t} =0.00065\ kg.m^{2}$
Friction coefficient	$f_{t} =0.017\ N.m/sec$
Air density:	$\rho=1.22\ Kg/m^{3}$

Table. A.3 The simulation conditions using Matlab/Simulink® R2009a

Type:	Fixed-step.
Ode-4:	Range Kutta Order4
Fixed-step Size (Fundamental sample time):	Simulation studies: $1e^{-5}$.
Tasking mode for periodic sample time:	Auto.

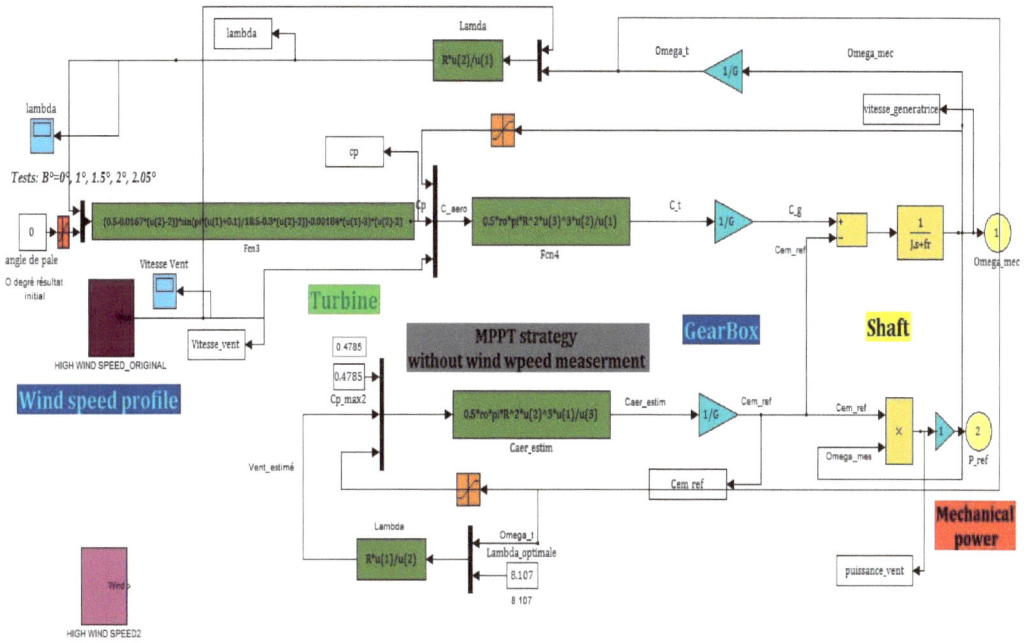

Fig. (A.1) Maximum power point tracking (MPPT) strategy of wind-turbine DFIG.

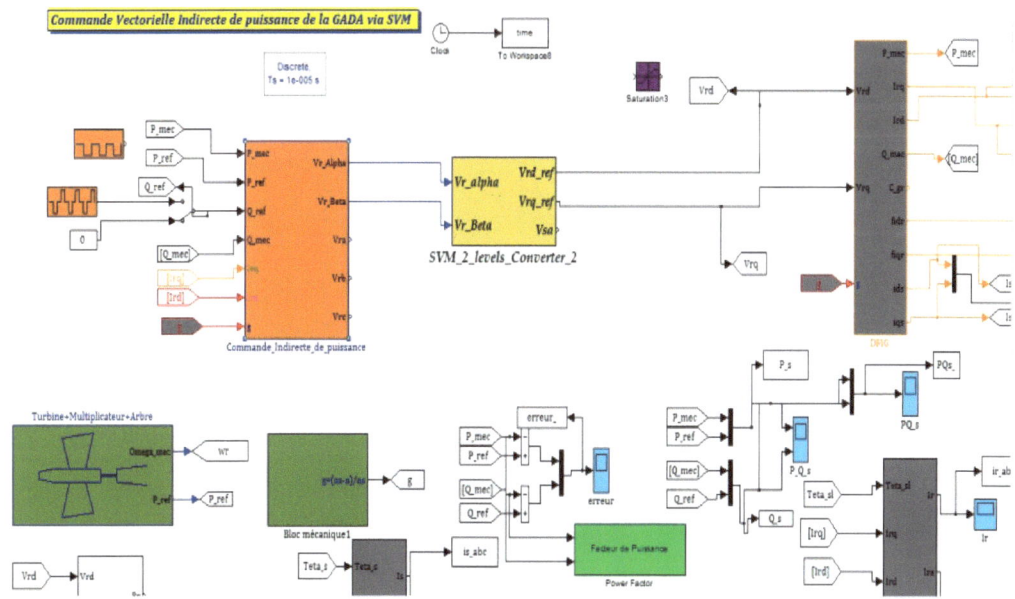

Fig. (A.2) Wind energy conversion system using DFIG (MATLAB/Simulink interface).

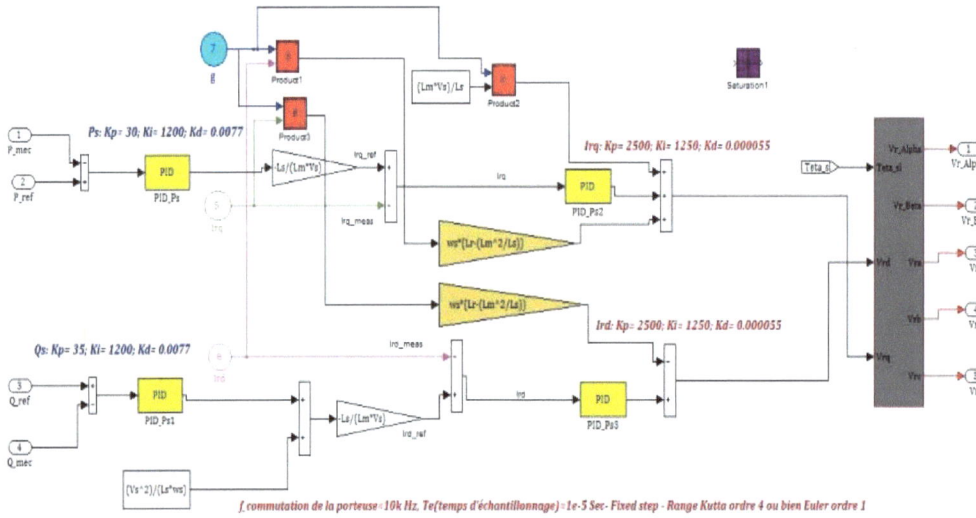

Fig. (A.3) Indirect power control (IDPC) for DFIG under MATLAB/Simulink interface.

Table. A.4 The gains's values of the proposed controllers.

	PI	PID	MRAC	T1-FLC	T2-FLC	NFC
Chapter 3:	P_s and Q_s K_p=30 K_i=1200 I_{rq} and I_{rd} K_p=2500 K_i=1250	P_s and Q_s K_p = 30 K_i = 1200 K_d = 0.0077 I_{rq} and I_{rd} K_p = 2500 K_i = 1250 K_d = 0.000055	I_{rq} and I_{rd} $K=5e^{+5}$. K_α=0.55. K_β=0.0002. K_{G1}=0.55. K_{G2}=0.0001.	-	-	-
Chapter 4:		P_s and Q_s K_p = 41 K_i = 2000 K_d = 0.007099		I_{rq} and I_{rd} G_1 = 1/2000 G_2 = 1/2000 G_3 = 2000	I_{rq} and I_{rd} G_1 = 1/2000 G_2 = 1/2000 G_3 = 2000	I_{rq} and I_{rd} G_1 = 1/2500 G_2 = 1/2500 G_3 = 1500

LIST OF ABBREVIATIONS

DFIG   Doubly Fed Induction Generator.

WECS   Wind Energy Conversion System.

VS-WECS   Variable Speed-Wind Energy Conversion System.

DFIM   Doubly Fed Induction Machine.

T1-FLC   Type1-Fuzzy Logic Control.

T2-FLC   Type2-Fuzzy Logic Control.

FIS   Fuzzy Inference System.

ANFIS   Adaptive Neuro-Fuzzy Inference System.

NFC   Neuro Fuzzy Control.

PI   Proportional Integral.

PID   Proportional Integral Derivate.

PWM   Pulse Width Modulation.

SVM   Space Vector Modulation.

AC   Alternating Current.

DC   Direct Current.

ADC   Analog to Digital Converter.

DAC   Digital to Analog Converter.

MPPT   Maximum Power Point Tracking.

P-DPC   Predictive-Direct Power Control.

IDPC   InDirect Power Control.

FOC   Field Oriented Control.

VOC   Vector Oriented Control.

SMC   Sliding Mode Control.

I/OLDC   Input/output Linearizing and Decoupling Control.

FFT   Fast Fourier Transform.

IGBT   Insulated Gate Bipolar Transistor.

DCM   Direct Current Motor.

IM   Induction Motor.

NPC   Neutral Point Clamped.

MRAC   Model Reference Adaptive Control (Controller).

LAS   Automatic Laboratory of Setif.

MRAS   Model Reference Adaptive System (Observer).

PF Power Factor.

MATLAB Matrix Laboratory.
     ®

THD Total Harmonic Distorsion.

Rpm Revolution per minutes.

PLL Phase locked loop.

ANN Artificial neural network

GSC Grid side converter.

RSC Rotor side converter.

Non-linear programming.

Non-dominated solving genetic algorithm II.

Outer approximation.

Parametric complementary pivot.

Probability density function.

Parametric mixed-integer linear programming.

Quadratic programming.

Robust optimization.

Rational reaction.

Simulated annealing.

Single objective optimization.

Single objective optimization programming.

Srinivas test function.

Triangular fuzzy number.

Journal of water resource and hydraulic engineering.

Water resource system.

List of Acronyms

B Degree (°). Blade pitch angle.

TSR - Tip Speed Ratio or λ (Lambda).

C_p - Power coefficient.

ρ (Kg/m3). Air density.

v (m/sec). Wind speed.

R_s (Ω). Stator Resistance.

R_r (Ω). Rotor Resistance.

I_{sq} (A). Stator quadrature (or transversal) current component.

I_{sd} (A). Direct axis current of stator.

I_{rq}^* & I_{rd}^* (A). Rotor reference quadrature (or transversal) and direct current components.

I_{rq_meas} & I_{rq_meas} (A). Rotor measured quadrature (or transversal) and direct current components.

L_s (H). Stator inductance.

L_r (H). Rotor inductance.

L_m (H). Magnetizing (mutual) inductance.

Φ_{rq} & Φ_{rd} (Wb). Rotor quadrature (or transversal) and direct flux components.

Φ_{sq} & Φ_{sd} (Wb). Stator quadrature (or transversal) and direct flux components.

V_{rq} & V_{rd} (V). Rotor quadrature (or transversal) and direct voltage components.

V_{sq} & V_{sd} (V). Stator quadrature (or transversal) and direct voltage components.

P_s, P_r & P_m (W). Stator power, Rotor power and Mechanical power.

P_s^* & Q_s^* (W & Var). Stator reference active power and Stator reference reactive power.

P_{s_meas} & Q_{s_meas} (W & Var). Stator measured active power and Stator measured reactive power.

T_{em}, T_{aero} & $T_{gearbox}$ **(N.m).**   Electromagnetic torque, Aerodynamic torque and Gearbox torque.

T_r & T_{vis} **(N.m).**   Load torque and Viscous torque.

J_G, $J_{Turbine}$ & **J** **(Kg.m2).**   Generator moment inertia, Turbine inertia and Total inertia.

P - Number of Pole pairs.

Ω_{mec} **(Rad/sec).**   Mechanical speed.

N_r **(rpm).**   Rotor speed.

$N_{synchronous}$ **(rpm).**   Synchronous speed (=1500 (rpm)).

G - The gain of the gearbox.

f_{DFIG} **(N.m/sec).**   DFIG's Friction coefficient.

$f_{Turbine}$ **(N.m/sec).**   Turbine's Friction coefficient.

S - Slip.

ω_s & ω_r **(rad/sec).**   Stator and rotor pulsations.

θ_s, θ_r & θ_{slip} **(rad).**   Stator', rotor' and slip angles.

V_{dc} **(V).**   DC-Link voltage.

C_{dc} **(F).**   DC-capacitance

f_{grid}, f_{stator} & f_{rotor} **(Hz).**   Grid, stator and rotor frequencies.

I_{s_abc} & I_{r_abc} **(A).**   Stator and rotor currents (line to neutral).

V_{s_abc} & V_{r_abc} **(V).**   Stator and rotor voltages (line to neutral).

K_P, K_I & K_D - Proportional gain, Integral gain and Derivative gain.

I_{sa_meas} & I_{ra_meas} **(A).**   Stator and rotor measured current (line to neutral).

S_w **(m²).**   Wind turbine blades swept area (= $\pi*R2$).

R (m).   Blades diameter of the turbine.

$\mathbf{I_{g\_abc}}$ **&** $\mathbf{I_{g\_abc}}$ ***** **(A).** Grid measured and reference currents (line to neutral).

$\mathbf{I_{rec}}$ **(A).** Rectifier current.

$\mathbf{C_{p\_max}}$ **&** $\mathbf{\lambda_{\_optimal}}$ - Maximum power coefficient and optimal Tip speed ratio (TSR or λ).

$\mathbf{T_{ij}}$ **&** $\mathbf{K_{ij}}$ - IGBTs of the rectifier and inverter respectively (i=1, 2 'lines' & j=1, 2, 3 'columns').

$\mathbf{S_{ij}}$ - IGBT variable states (1 or 0).

$\mathbf{C_f}$ **&** $\mathbf{L_f}$ **(F and H).** Filter capacitance and filter inductance.

$\mathbf{I_{fa}}$, $\mathbf{I_{fb}}$ **&** $\mathbf{I_{fc}}$ **(A).** Filter currents (in each phase).

$\mathbf{V_{fab}}$, $\mathbf{V_{fbc}}$ **and** $\mathbf{V_{fca}}$ **(V).** Filter voltages (line to line).

$\mathbf{L_{f^*h(x)}}$ - Lie derivative.

$\mathbf{T_1}$, $\mathbf{T_2}$ **&** $\mathbf{T_0}$ **(sec).** The SVM switching time duration.

$\mathbf{\mu_4}$, $\mathbf{\mu_3}$, $\mathbf{\mu_2}$ **&** $\mathbf{\mu_1}$ - The NFC's learning-rate parameters.

SUBJECT INDEX

A

Absence of reactive power compensation 13
Active power error 106, 107, 111
   T1FLC 106, 107, 111
   T2FLC 106, 107, 111
Active power point 68, 105
Adaptive 3, 17, 63, 87
…disturbance rejection control (ADRC) 17
   gains 63
   neuro-fuzzy inference system (ANFIS) 3, 87
Artificial neural networks (ANNs) 3, 5, 88
Asynchronous and synchronous generator 9

B

Backstepping 3, 18
   control (BSC) 18
   mode control (BMC) 3
Behavior 68, 74, 77, 80, 81, 115
   of stator 80, 81, 115
   of wind-system parameters 74
   of wind-system parameters of mode 68, 77
Bench, experimental test 17, 122
Big 26, 73, 75, 77, 78, 115
   power error 26, 73, 77
   undulations 75, 78, 115

C

Changes direction 56, 57
Closed loop transfer function (CLTF) 36, 37
Color, green 5, 68, 80, 82, 105, 115, 119
Comparative study 15, 26, 86, 88
Complex nonlinear systems 96, 97
Components, axes rotor currents 5
Control 18, 28, 45, 52, 63
   adaptive 28, 63
   backstepping 18
   DFIG, indirect power 45, 52
Controller 5, 17, 43, 60, 62, 63, 82, 89, 101, 102, 121
   adaptive 17, 62
   conventional 5, 82

hysteresis 43
input 60
neuro-fuzzy 101, 102
robust 5, 121
signal 60, 62, 63
stator powers 82
topology 89
Control rotor 64, 88, 96, 100, 105
   quadrature 88
Control strategies 3, 9, 15, 16, 17, 18, 21, 43
   vector 43
Control structure for power electronics
   converter 14
Control system 13, 62, 91
   adaptive 62
Converters, back-to-back PWM 41, 42
Curves/robustness tests section 5

D

Decoupling control 19
Defuzzification 89, 90, 92, 93, 94, 100, 103
Derivative terms (DTs) 15
DFIG 3, 13, 18, 19, 27, 61, 80, 81, 87, 115, 116, 121
   -based wind system 3
   -based wind turbine-generator systems 13
   power control of 19, 121
   rated power of 80, 81, 115, 116
   wind turbine 18, 61
   wind-turbine 27, 87
Doubly fed induction machine (DFIM) 1, 15, 56, 57, 58, 60

E

Electromotive forces 32
Equations, first-order differential 54
Excellent THD of stator currents 74, 75, 77, 78, 112, 114
Expressions, rotor flux 34
Extended Kalman filter (EKF) 17

F

Fault ride through (FRT) 12
Field oriented control (FOC) 2, 16, 17
Functional model predictive control (FMPC) 19
Fuzzy 5, 87, 91, 92, 94, 96, 99, 100, 101
   controllers 5, 92, 99
   rule base 94, 96, 100, 101
   system 87, 91, 94
   type 87
Fuzzy sets 87, 90, 91, 92, 93, 96, 97, 98, 99, 103
   operations 90, 91
   model 87
   theory 90, 91
Fuzzy logic 5, 16, 20, 86, 87, 88, 89, 90, 91, 92, 93, 94, 95, 96, 97, 98, 99, 100, 103, 104, 117
   control (FLCs) 5, 16, 20, 86, 87, 88, 89, 92, 96, 97, 98, 103, 104, 117
   controller 87, 89, 91, 92, 93, 94, 96, 98, 100
   rules 91, 92
   surface 94, 95
   systems (FLSs) 87, 90, 96, 97, 98, 99
   toolbox 97, 98

G

Gaussian functions 103, 105
Gearbox 2, 12, 89
Generator 2, 41, 59
   mode 41, 59
   parameters changement 2
   rotor speed/position 2
   type 2
Generation system 9, 13
   based power 9
Genetic algorithm (GA) 2, 122
Global 10, 89
   increasing wind power capacity 10
   wind-turbine system scheme 89
Good tracking power 73, 75, 76, 79, 108, 109, 112, 114
Grid 21, 26, 28, 41, 42, 43, 44, 89
   -connection mode 26

side converter (GSC) 21, 26, 28, 41, 42, 43, 44, 89
GSC configuration 42, 43

H

High wind-power generation (HWPG) 4, 88

I

IDPC with/without robustness tests 28
Improved IDPC 112, 121
Induction generator (IG) 1, 2, 9, 12, 27, 56, 86
   doubly-fed 12, 27
   wound rotor 9, 12
   wound-rotor 2
Input 3, 91, 93, 98, 102, 103
   crisp 91, 93, 98
   fuzzy sets 93, 98
   layer 102, 103
   -output feedback linearization control 3
   -1output Membership function editor of type- 98
Installed wind power production 1
Integration, efficient wind power 3
Intelligent controllers, artificial 118, 122
Inverter voltage vectors 46, 47

K

Kirchhoff's voltage law, applying 54, 55
Knowledge base 91, 92, 98

L

LC Filter 52, 89
Learning algorithm, on-line 101, 104
Line voltage 47, 48
Load variation 122
Loops, parameter adjustment 62
Low-depth voltage dips (LDVD) 17

M

Maximum power point tracking (MPPT) 2, 4, 15, 20, 26, 28, 39, 40, 66, 67, 70, 72, 73, 74, 106, 107, 111, 116, 117

Measured powers 73, 74, 75, 76, 77, 78, 79, 108, 109, 112, 114

Membership functions (MFs) 91, 92, 93, 94, 95, 96, 97, 98, 99, 101, 102, 103
 editor of Type- 97
 secondary 98, 99
 shape of 91

Model, wind-turbine 26

Model predictive control (MPC) 3, 19, 21

Model reference adaptive control (MRAC) 17, 26, 28, 63, 64, 66, 67, 68, 70, 72, 73, 74, 75, 76, 77, 78, 79, 80, 81, 117, 122

Model reference adaptive system (MRAS) 17, 28

MPPT strategy 26, 74, 77, 105, 109
 for novel IDPC algorithms 105
 - Random Wind Speed 77
 - Step Wind Speed 74, 109
 with/without robustness tests 26

MRAC 67, 70
 active power error 67, 70
 reactive power error 67, 70

N

Network 4, 11, 12, 13, 17, 18, 19, 30, 33, 101, 102
 electrical 17, 18, 30, 33
 neuro-fuzzy 101, 102

Neural networks (NN) 3, 87

Neuro-fuzzy 5, 61, 86, 88, 89, 101, 102, 105, 106, 107, 108, 109, 110, 111, 112, 113, 114, 115, 116, 117, 118, 121, 122
 control (NFC) 5, 61, 86, 88, 89, 101, 102, 105, 106, 107, 108, 109, 110, 111, 112, 113, 114, 115, 116, 117, 118, 121, 122
 logic 86

Neutral Voltage Line 47, 48

NFC 101, 105, 106, 107, 109, 111, 112, 114, 117
 active power error 106, 107, 111
 controller 101, 105, 109, 112, 114, 117
 of rotor 101
 reactive power error 107, 111

Novel type regulator approaches 20

O

Open loop transfer function (OLTF) 37

Operation 13, 16 56, 57, 59, 75, 78, 91
 generating 56
 super-synchronous 16, 56, 59, 75, 78

Operational modes 56

Operator skill 93

Order 15, 17, 18
 models, reduced 15
 sliding-mode control 18
 sliding mode supervisor, advanced 17

Output 12, 16, 46, 47, 60, 93, 96, 98, 122
 control signal 60
 fuzzy sets 93, 98
 power 12, 122
 power generator 16
 processor 96
 voltages 46, 47

Overshoot 26, 73, 74, 75, 77, 80, 81, 115, 116, 121
 big 77, 80, 81, 115, 121
 high 81, 116
 lower 81, 115
 neglected 73, 75, 77
 remarkable 80
 remarkable 26, 74, 80

P

Parameter 17, 82, 119
 changement 17, 82, 119

Parameters variation 5, 26, 65, 79, 80, 86, 88, 93, 96, 100, 105, 106, 114, 115, 122

Particle swarm optimization (PSO) 2, 122

Performances 19, 86

high Wind-turbine 19
wind system 86
PI 5, 18, 19, 26, 27, 28, 36, 44, 73, 74, 80, 81, 82
 and PID controllers 26, 82
 compensator 18, 19
 controllers 5, 26, 27, 28, 36, 44, 73, 74, 80, 81
PID 67, 70
 active power error 67, 70
 reactive power error 67, 70
PID controllers 5, 26, 59, 60, 64, 73, 75, 76,
  79, 81, 82, 86, 87, 88, 89, 96, 100, 105,
  118, 121
 classical 121
 conventional 87
Pitch 13, 14, 15, 19, 39, 40, 69, 71, 75, 78,
  110, 113
 angle control 15, 19
 angles 13, 14, 39, 40, 69, 71, 75, 78, 110,
  113
PI to control stator powers and rotor currents
  82
Power 1, 2, 3, 5, 9, 10, 14, 44, 56, 57, 65, 68,
  69, 72, 89, 90, 114, 122
 active 2, 4, 5, 12, 13, 14, 15, 16, 17, 18, 19,
  28, 43, 56, 57, 58, 66, 67, 68, 69, 70, 71,
  72, 73, 74, 75, 77, 78, 80, 81, 88, 105,
  106, 107, 110, 111, 112, 113, 114, 115,
  116, 117
 converter 2, 12, 15, 56, 58
 electronics (PE) 1, 2, 9, 10, 14
 generated 13
 measured active 14, 81, 116
 rated 4, 12, 88
 wind generator 57
Power capacity 2, 9
 increasing wind 9
Power coefficient 26, 38, 39, 40, 69, 71, 75,
  78, 110, 113
Power control 3, 4, 15, 16, 17, 18, 19, 26, 27,
  30 36, 88, 89, 96, 101, 121
 active 15, 19
 conventional 4, 88
 conventional Indirect 27, 30, 36
 direct 3, 16, 17, 18

Power error 66, 67, 70, 75, 77, 78, 80, 106,
  107, 108, 109, 111, 115
 lower 77, 78, 80, 115
 reactive 66, 67, 70, 106, 107, 111
 small 75, 80, 108, 109, 115
 value of 80, 115
Power factor (PF) 4, 19, 44, 65, 66, 67, 69, 70,
  72, 74, 77, 80, 81, 106, 107, 111, 118,
  122
 PF NFC 106, 107, 111
Power flow 13, 26, 56, 58, 87
 diagram 56, 58
Power tracking 17, 26, 28, 74, 76, 77, 78, 79,
  82, 109, 112, 114, 117, 118, 121
 excellent 77, 79, 112, 114
 good 76, 77, 79, 109, 114
 good maximum wind 17
 poor 74, 76, 77, 78, 79, 109
Predictive 3, 16, 19, 122
 control 3, 16, 122
 direct power control (PDPC) 19
Proportional-integral (PI) 4, 18, 20, 26, 27, 28,
  36, 59, 61, 66, 67, 68, 69, 70, 71, 72, 73,
  74, 75, 76, 77, 78, 79, 80, 81, 82, 89,
  117, 122
 -derivate (PID) 4, 26, 28, 61, 66, 67, 68, 70,
  72, 73, 74, 75, 76, 77, 78, 79, 80, 89,
  117, 122
Pulse width modulation (PWM) 21, 28, 43, 46

Q

Quadrature 64, 66, 67, 68, 70, 74, 77, 86, 96,
  100, 105, 106, 107, 111, 121
 currents 64, 66, 67, 68, 70, 74, 77, 86, 96,
  100, 105, 106, 107, 111, 121
 fluxes 66, 67, 70, 106, 107, 111

R

Random wind speed 40, 68, 70, 73, 81, 105,
  111, 112, 113, 117, 118
Reactive powers 3, 4, 5, 13, 17, 26, 27, 28, 29,
  33, 36, 43, 44, 60, 66, 67, 68, 69, 70, 71,

72, 73, 74, 75, 77, 78, 79, 80, 81, 86,
    105, 106, 107, 108, 109, 110, 111, 112,
    113, 114, 115, 116, 117, 118, 122
compensation 13
error PI 67, 70
error T1FLC 106, 107, 111
error T2FLC 106, 107, 111
references 71, 107
tuning 5
variations 27
Recapitulation 74, 76, 78
Reference 28, 39, 44
    model output 28
    stator power 39
    transversal grid 44
Regulators 2, 4, 5, 19, 33, 34, 36, 88, 89
    conventional 4, 36, 88
    intelligent 5, 89
    of rotor currents 33
    of stator powers 34
Remarkable 1, 4, 9, 80, 81, 88, 115
    power error 4, 80, 81, 88, 115
    energy 1, 9
Resistances, global rotor 12
Ripples, remarkable 76, 79
Robustness tests 4, 5, 26, 61, 68, 72, 73, 79,
    80, 82, 86, 88, 106, 114, 115, 116, 117,
    118, 119, 121
    of mode 72, 73, 116, 117
Rotor 3, 14, 16, 17, 18, 21, 26, 28, 29, 31, 32,
    34, 41, 44, 46, 56, 57, 59, 93, 118
    measured 59, 93
    angle 44, 59
    blade 14
    circuit 56
    fluxes 3, 29, 31, 34
    flux estimation 17
    flux vectors 31
    position 17
    power 56, 57
    quadrature 93, 118
    side converter (RSC) 16, 18, 21, 26, 28, 41,
        44, 46, 93
    sinusoidal 59

voltages and rotor currents 32, 34
Rotor currents 4, 32, 33, 34, 35, 44, 66, 67, 69,
    70, 71, 72, 74, 75, 76, 77, 79, 81, 82, 106, 107,
    109, 110, 111, 112, 113, 114, 118, 122
    axes 44
    behavior 76, 79
    control 122
    desired 35
    phase 35
    sinusoidal waveforms of 76, 79
Rotor resistance 2
    control 2
Rotor voltages 28, 29, 31, 32, 34, 35, 54
    line 54
    two-phase 34
Rules 5, 90, 92, 95, 96, 100, 118
    fuzzy control 95, 100
    linguistic 5, 90, 96, 100, 118
    linguistic control 92

S

Simpower systems 6
Simulation study, comparative 17, 28, 68, 105,
    106
Sinusoidal form 69, 72, 74, 75, 77, 78, 109,
    112
Sliding mode control (SMC) 3, 18
Space vector 26, 27, 28, 46, 47, 49, 50, 51, 52,
    81, 87, 118
    modulation (SVM) 26, 27, 28, 46, 51, 81,
        87, 118
    PWM (SVPWM) 47, 49, 50, 51, 52
Speed 4, 5, 15, 17, 26, 56, 57, 59, 62, 68, 74,
    75, 76, 78, 79, 81, 82, 86, 89, 106, 118,
    121
    mode control 15
    range 56, 57
    variation 4, 5, 17, 26, 59, 62, 68, 74, 75, 76,
        78, 79, 81, 82, 86, 89, 106, 118, 121
Squirrel-cage induction generator (SCIG) 11,
    16
Stator
    reference 74, 77

active power 71
  and rotor flux vectors 31
  angle 44
  currents 30, 31, 34, 66, 67, 69, 70, 72, 74,
    75, 77, 78, 106, 107, 108, 109, 111, 112,
    114
Stator flux 12, 17, 19, 16, 26, 30, 32, 33, 71,
    72, 73, 116, 117
  estimated 17
  orientation 19, 26
  pulsation ωs 33
  Reactive Power 71
  reactive 17
  reactive power 72, 73, 116, 117
  resistance 32
  vector 30
  windings 12
Stator powers 4, 33, 34, 56, 81, 122
  and rotor currents 33
  control independely DFIG's 81
Stator voltages 30, 33, 122
  control 122
Steady states 4, 5, 19, 28, 59, 68, 73, 74, 75,
    76, 77, 78, 79, 80, 81, 82, 86, 106, 108,
    109, 112, 114, 115, 116, 117, 118
Sub-synchronous 16, 56, 57, 58, 59, 68, 75,
    78, 82
  mode 59
  motoring mode 57, 58
  generating modes 57, 58
  modes 59, 68
  motoring mode 57, 58
  speeds 56, 57, 82
Switching frequency 26, 27, 28, 49, 86
  fixed 26, 28, 86

T

Tgearbox 29, 41
Tip speed ratio (TSR) 38, 39, 40
Topology 1, 12, 41, 42, 45, 52, 68, 74, 77, 80,
    81, 105, 114, 115, 116, 117, 118
  circuit 41, 42
  rotor side converter 45, 52

Torque, electromagnetic 15, 29, 32, 39, 59
Tracking, maximum power point 2, 4, 26, 28,
    39, 40
Transfer function, closed loop 36, 37
Transient 4, 5, 19, 26 28, 59, 68, 72, 73, 74,
    75, 76, 77, 78, 79, 80, 81, 82, 86, 106,
    108, 109, 112, 114, 115, 116, 117, 118
  and steady states 4, 5, 19, 28, 59, 68, 73, 74,
    75, 76, 77, 78, 79, 80, 81, 82, 86, 106,
    108, 109, 112, 114, 115, 116, 118
  state 26, 72, 73, 116, 117
Type 15, 16, 87, 96, 97, 98
  brushless 15
  classical 96
  numerous 16
  ordinary 98
  reducer 96
  reducing 97
  traditional 87

U

Uncertainties, dynamic 87
Undulations 80, 81
  remarkable 80, 81
Unity power factor 19, 44, 65, 74, 80, 81, 115,
    117, 118

V

Variable 12, 13
  rotor resistance 12
  -speed wind turbine 13
Variation, wind-speed 86, 118, 121, 122
Vectors, switching variable 46
Voltage 17, 28, 29, 31, 47, 48, 54, 55
  dips 17
  equations, following 54, 55
  stator and rotor 28, 29, 31
  vectors switching vectors line 47, 48

W

Wind energy 27, 99
  conversion system 27
  systems 99
Windings 12
  rotor phase 12
  stator phase 12
Wind power (WP) 9, 11, 14, 16, 27, 28, 75,
    109, 112
  applications (WPAs) 9
  captured 16
  converter 14
  development 9
   generation (WPG) 9, 11, 27
Wind speeds 11, 17, 68, 105
  low 11, 68, 105
  time-varying 17
Wind-system 5, 28, 68, 74, 77, 105, 117, 118,
    121, 122

in terms 28, 121
mathematical 5
parameters 68, 74, 77, 105
performance algorithms 86
performances recapitulation 117
performances regardless, improved 118
variation 122
Wind turbine (WT) 1, 2, 9, 11, 12, 13, 16, 17,
    18, 26, 27, 37, 38, 41, 56, 61, 65, 66, 82,
    105, 119
  concept 9, 11
  -DFIG algorithm 65
  -generator systems, based 9
  mathematical model 38
  measurements 16
  model 41
  parameters 119
  power conversion efficiency 38
Wound-rotor induction generator (WRIG) 2,
    9, 12